The University of Michigan
Center for Chinese Studies

Science, Medicine, and Technology in East Asia

Volume 3

Nathan Sivin, General Editor

The University of Michigan
Center for Chinese Studies

Science, Medicine, and Technology in East Asia

Volume 2

Nathan Sivin, General Editor

Science and Medicine in Twentieth-Century China: Research and Education

edited by
John Z. Bowers
J. William Hess
Nathan Sivin

Ann Arbor

Center for Chinese Studies
The University of Michigan

1988

Copyright © 1988
by
Center for Chinese Studies
The University of Michigan
All Rights Reserved

Library of Congress Cataloguing-in-Publication Data:

Science and medicine in twentieth-century China: research and
education / edited by John Z. Bowers, J. William Hess, Nathan Sivin.
    p.   cm. -- (Science, medicine, and technology in East Asia:
v. 3)
    Essays originated at a conference held at the Rockefeller Archive
Center, Pocantico Hills, Tarrytown, N.Y., 21-23 May 1984.
    Includes bibliographies and index.
    ISBN 978-0-89264-077-5 - ISBN 978-0-89264-078-2 (pbk)
    1. Medical policy--China--Congresses. 2. Science and state-
-China--Congresses. I. Bowers, John Z., 1913-   . II. Hess, J.
William. III. Sivin, Nathan. IV. Series.
    [DNLM: 1. Education, Medical--history--China--congresses.
2. History of Medicine, 20th Cent.--China--congresses. 3. Public
Policy--history--China--congresses. 4. Research--history--China-
-congresses. 5. Science--history--China--congresses. W1 SC744 v.
3 / Q 158.5 S4156 1984]
RA395.C53S35   1988
338.95106--dc19
DNLM/DLC
for Library of Congress   88-36952
                                                                                                                                                       CIP

Cover design by Peter Sasgen

Cover: "The Results of Foot Binding: An X-Ray Photograph of a Chinese Woman's Foot," a sketch made from a photograph by the Rev. W. A. Cornaby, published in *The Illustrated London News* for 15 February 1908. The Chinese "proverb" in the upper right is translated there: "To tamper with creation is to spoil the harmony of heaven."

Printed and bound by CPI Group (UK) Ltd, Croydon, CR0 4YY

6 5 4 3 2 1

Contents

Science, Medicine, and Technology in East Asia, *vii*
Contributors, *ix*
Introduction, *xi*
  —Nathan Sivin

Part I: Republican China

Genetics in Republican China, *3*
  —Laurence A. Schneider

Botany in Republican China: The Leading Role of Taxonomy, *31*
  —William J. Haas

The Development of Geology in Republican China, 1912–1937, *65*
  —Tsui-hua Yang

Comment on the Historical Essays, *91*
  —Ka-che Yip

Part II: People's Republic of China

The Role of Biomedical Research in China's Population Policy, *99*
  —Sheldon J. Segal

"Farewell to the Plague Spirit": Chairman Mao's Crusade against Schistosomiasis, *123*
  —Kenneth S. Warren

Human Viral Vaccines in China, *141*
  —Scott B. Halstead and Yu Yong-Xin

Genetics in Postwar China, *155*
—James F. Crow

Agriculture and Plant Protection in China to 1980, *171*
—Robert L. Metcalf

Food Policy and Nutritional Status in China, 1949–1982, *201*
—Dean T. Jamison and Alan Piazza

World Bank Experience in Education in China, 1980–1984, *233*
—Frank Farner

Medical Education in China, *239*
—John R. Evans

Index, *261*

Science, Medicine, and Technology in East Asia

The Series is sponsored by *Chinese Science* and the Center for Chinese Studies at The University of Michigan. The Editorial Board includes:

Hans Ågren, M.D.
    University of Uppsala, Sweden (medicine)
Professor Kenneth J. DeWoskin
    The University of Michigan, U.S.A. (literary aspects)
Dr. John S. Major
    The Asia Society, New York, U.S.A. (early science and technology)
Dr. Shigeru Nakayama
    Tokyo University, Japan (general)
Joseph Needham, F.R.S., F.B.A.
    East Asian History of Science Library, Cambridge, England (general)
Professor Laurence A. Schneider
    State University of New York at Buffalo, U.S.A. (twentieth-century science)
Professor Nathan Sivin
    University of Pennsylvania, U.S.A. (Chairman; general)
Professor E-tu Zen Sun
    Pennsylvania State University, U.S.A. (technology)

The Series publishes books on traditional and modern science, medicine, and technology in China, Japan, and Korea, based on research in primary sources in the languages of those societies or on the study of artifacts. It applies the highest standards of refereeing, technical editing, and production to books which combine scientific and East Asian content. Its aim is to produce books of the highest scholarly quality at prices nonspecialists can afford.

Orders and inquiries about the availability of titles in print should be addressed to: *Publications,* Center for Chinese Studies, 104 Lane Hall, The University of Michigan, Ann Arbor, MI 48109.

Contributors

Authors

James F. Crow, Professor of Genetics, University of Wisconsin

John R. Evans, Chairman, Allelix, Inc.

Frank Farner, Principal Educator, East Asia and Pacific Education Region, The World Bank

William J. Haas, Assistant Professor of History, Washington University in St. Louis

Scott B. Halstead, Acting Director, Health Sciences Division, The Rockefeller Foundation

Dean T. Jamison, Professor of Education and of Public Health, University of California at Los Angeles

Robert L. Metcalf, Professor Emeritus of Entomology, University of Illinois at Champaign-Urbana

Alan Piazza, Consultant, China Agriculture Operations Division, The World Bank

Laurence A. Schneider, Professor of Chinese History, State University of New York at Buffalo

Sheldon J. Segal, Director, Population Sciences Division, The Rockefeller Foundation

Kenneth S. Warren, Associate Vice President, Molecular Biology and Information Sciences, The Rockefeller Foundation

Tsui-hua Yang, Institute of Modern History, Academia Sinica, Taiwan

Ka-che Yip, Associate Professor of History, University of Maryland–Baltimore County

Yu Yong-Xin, Deputy Director, National Institute for the Control of Pharmaceutical and Biological Products, Ministry of Health, Beijing, China

Editors

John Z. Bowers, Former President, Josiah Macy, Jr., Foundation

J. William Hess, Former Associate Director, Rockefeller Archive Center

Nathan Sivin, Professor of Chinese Culture and of the History of Science, University of Pennsylvania

Introduction

Nathan Sivin

The essays in this book originated at a conference held at the Rockefeller Archive Center, Pocantico Hills, Tarrytown, New York, 21–23 May 1984. It brought together a group of people engaged in studies of twentieth-century China from the perspective of science, medicine, or technical history. Their topics were not prescribed. There is so little well-informed scholarship in this area that the point was simply to elicit papers that may help to define a research frontier. The result seemed, despite the diversity of topics, potentially useful to many readers. Hence this book, which offers in revised form a selection of the original essays.

These essays fall naturally into two groups. Eight grow out of the personal experience of scientists who in one way or another have been deeply involved in China's recent efforts to modernize. They are writing contemporary history, in most cases with explicit attention to a longer or shorter span of the past. Three papers by historians look synoptically and analytically at attempts to root modern science in China during the Republican period (1912–1949).

Accounts of Contemporary China

Most of what we know about science and medicine in China today comes either from accounts of an expert's first visit to China, usually for three weeks, or from Chinese publications, which tend to be semi-official in character, informative but unrevealing. Some American scientists know a great deal about what is happening in China because they help make it happen. They work for the World Bank, World Health Organization, Rockefeller Foundation, or some similar organization that cooperates with the Chinese government or Chinese institutions to shape change. Accounts of their experience usually appear in institutional publications meant for limited distribution.

These international scientists have a great deal to say, not only about the whats and whys of science policy, but about how it is working. Robert L. Metcalf provides an overview of changes in Chinese agriculture in the People's Republic, with particular attention to protecting plants from the pests that compete with human beings for them. He notes that in Southeast Asia the dependence on insecticides, each of which eventually loses its effectiveness, is leading to catastrophe, and contrasts with this shortsightedness the integrated Chinese approach that begins with ecological studies and minimizes the use of insecticides. In certain circumstances weeding may even be reduced to let spiders multiply that prey on planthoppers. At the same time he reminds us that attempts to copy the Soviet model of large-scale mechanized farming in the 1950s wasted Chinese resources, and that a permanent legacy of the Cultural Revolution era (1966–1976) was the shift to hand tractors and other flexible and affordable machinery. He also points out that more recent attempts to grow three crops per year wherever feasible have often increased expense without producing more food.

Kenneth S. Warren, in the best-informed study of schistosomiasis that I have seen, reexamines the decision formed over the years 1955–1958 to eradicate the disease at any cost. He points out that banishing schistosomiasis from Chinese soil was a choice from among three options; the other two were controlling infection and controlling cases in patients. He asks whether it was the best choice given the circumstances, and whether a different choice could have been more successful than an impractically ambitious policy about which Chinese scientists were given little say. His argument, subtle and commonsensical, results from his field studies in China of incidence and severity, and his awareness of the Japanese experience in doing away with schistosomiasis. It is a refreshing change from package-tour reports unaware that the disease still exists.

The other papers in this category are also neither rosy nor jaundiced. Frank Farner informally sketches a system of higher education expanding at dizzying velocity, and increasingly dependent on experiments never before tried on such a large scale. The enrollment of the television university, for instance, is already equal to that of "the whole system of residential colleges and universities." John R. Evans acknowledges the dedication to universal health care embodied in China's effort to expand medical education, but sees that human and other resources will be wasted so long as planners let themselves be pushed by technology rather than pulled by patterns of need. Dean T. Jamison and Alan Piazza find that, despite large advances in nutrition, the game is not yet won.

James F. Crow reflects on the lasting damage done to genetics research first by the hegemony of Lysenkoism (absolute up to 1956 and powerful long afterward), and then by the Cultural Revolution. Sheldon J. Segal appraises population policy from the standpoint of how it uses biomedical research. He finds that the crucial research questions have ceased to be those of finding methods people will use, or finding effective ways to limit the number of children per family. Despite early missteps the Chinese have made these problems manageable, and have moved on to reducing side effects of contraception and lessening the need for abortion. Scott Halstead and Yu Yong-Xin see the Chinese approach to developing vaccines against viral diseases as a model for developing countries. Nations that buy their vaccines abroad rather than mastering production technologies find eventually that they lack the technical means to monitor their vaccination programs. At the same time, Halstead and Yu call attention to imprecision in Chinese field studies of efficacy and the difficulty of distributing vaccines in good condition and rigorously measuring their effectiveness in the field.

These essays reflect a viewpoint shared by many scientists who have been involved day to day with Chinese research and who see China neither as paradise nor hell. They see its path to science as individual, based to some extent on learning from countries that industrialized earlier but also offering some lessons from which the overdeveloped West can benefit. Americans can recognize in these assessments of change many of their own shortcomings—a habit of daydreaming about technological solutions for problems tractable only through reform or social change, a tendency to settle for the politically expedient policy even when it is the worst at hand, and a growing inability to reason about the human costs of "progress" that cannot be expressed in cold cash. Over the coming decades it will become clear whether Chinese and Americans can learn enough from each other's failures to overcome these deficiencies.

Finally, a word about a limitation of these studies. Naive readers may be misled, or may mislead themselves, into thinking that the breathless account of someone just back from a seventeen-day tour of eight Chinese cities is up-to-the-minute. That will no doubt be true of factual assertions about what the writer saw or did, but the web of significance that embeds these statements of fact is based on the conventional wisdom and the out-of-date statistics that are usually the only ones available for China. The authors of this volume may be less in thrall to platitudes, but their statistics, as they make clear, are as much as a decade old. Readers will do well to keep in mind that all accounts of contemporary China are best read—and read critically—as history.

Studies of Science in Republican China

When I was an undergraduate my teachers agreed that one could not write the history of events until they had matured for at least fifty years. Otherwise, they assured us, we wouldn't be far enough away from the hurly-burly to sort out such an overwhelming bulk of evidence, and couldn't be objective.

Remembering that lends a special zest to my present position, surrounded by colleagues and graduate students who use the craft of the historian to investigate science, technology, and medicine in the twentieth century. I see this work being done with care and rigor. I notice that it is now considered respectable. That leads me to wonder what anxieties of a generation ago made historians forget that mountains of documents may yield no evidence at all unless they are asking fruitful questions.

As for historical objectivity, following the work of a great many colleagues through a period of transition in methodology, contention about issues, and division over political assumptions has left me less than starry-eyed. When someone appeals to objective, value-free research, or to the certainty of scientific or quantitative method in history, I can't help asking myself what he is going to expect from me in exchange for his wooden nickel. Many of the great minds of our age from Marx on have tried to make history objective, value-free, and scientific, and ostentatiously failed. So long as history is about human choices—as it always is—and the best historians are concerned individuals living in bad times—as they always are—history is bound to remain a moral and aesthetic enterprise. Having taken in trade more than my share of wooden nickels, and having recirculated them from time to time, I have come to prefer authors who are aware of their values and frank about them over those who assure me that none enter their work. That is admittedly an aesthetic (and no doubt moral) preference.

These problems take on a special poignance with respect to the history of modern Chinese science, because we have two different Chinas to cope with (both of which include the Chinese mainland and Taiwan). Looking at the still meager publications on the Republican period, we find two frames of interpretation that hardly seem to portray the same society. One shows a China that succeeded brilliantly in science, the other a China that failed.

The first account sees the science of the prewar Republic as a march of progress. We find no more than an occasional aside about the chaotic, inefficient, and corrupt government of the period, the lack of public order that made people's lives desperate. We find little curiosity about the potential for self-sufficiency of a scientific establishment planned and paid for by foreigners. We ask ourselves how a country could hope to nurture scientific leaders when the great majority of the population outside a few cities was not being educated in the rational use of time and resources, and differed from their ancestors five hundred years earlier mainly in being materially worse off and politically more riven.

The second account sees Republican science as a failure. Peter Buck's recent book views the scientific leaders of the period as absurd figures in the "farce of Republican China replaying the tragedy, such as it was, of the social and scientific transformation of the United States in the late nineteenth and early twentieth centuries."[*] That is not the

---

[*] Peter Buck, *American Science and Modern China 1876–1936* (Cambridge, 1980), 231.

best-informed version of this interpretation, but it is typical in being a reaction to its starry-eyed predecessors.

Such interpretations may be more realistic than the chronicles of progress, but they are no more satisfactory as historical analysis. They largely ignore the question of just what Chinese scientists discovered and created against just what odds in just what circumstances. They ask instead why Chinese accomplishments do not compare with those of wealthier countries. They seem unaware that Western Europe and the United States were not just wealthier; well before the twentieth century their societies and mentalities had been transformed by familiarity with new sciences and habituation to new technologies. The verdict of failure is predictable, and not even very informative.

This picture of failure leaves us ill prepared to realize that Peter Buck's "comic" figures of the 1930s—those who survived, at least—continued to lead research groups in solid scientific work for decades without foreign support and in spite of isolation, poverty, and political harassment. Some of them built, managed, and rebuilt educational and research institutions. This interpretation does not help us explain how these same survivors, out of prison or back from the countryside a decade ago, are now planning and executing the most intense effort of scientific modernization in any poor country in the twentieth century.

We find celebrations of progress and litanies of failure equally unsatisfactory here, as they always are in historical writing for grown-ups. Both represent bygone fads in the history of science. In that profession the internalist approach, still conventional a generation ago, was largely devoted to chronicles of progress. The externalist approach, evolved in reaction to it in the late 1960s and strongly influenced by sociology, looked more critically and usually less charitably at its subjects. These tendencies linger on in recent Sinologists' studies of Republican science, but historians of science themselves have worn out both of these fashions, and have gradually been finding the middle ground where ideas and the communities that evolve them can be seen as aspects of the same process.

The papers presented in this book demonstrate that scholars of twentieth-century Chinese science are finding their way to this more sophisticated middle ground. Ultimately, as they feel more at home there, we can expect that less space in published writings will be devoted to whether one scientific field or another succeeded or failed, or to whether the foundations who helped train the scientists were angels or devils. There are more intellectually substantial, more subtle problems that command our attention. These essays, among their many merits, suggest several such problems and give us food for thought about them. These questions in particular ought to be explored:

First, what accounts for the Nanking government's policy of support of sciences? We are not going to make much headway if we assume it was simply because all enlightened people are in favor of science: The "Science and Life-view" (*k'o-hsüeh yü jen-sheng-kuan* 科學與人生觀) polemic that began in 1923 made it clear that science and liberal democracy were not universally accepted as the salvation of China even among Western-educated intellectuals or politicians. Outside the universities the picture was no clearer by the outbreak of the Sino-Japanese War. The "red versus expert" debate is not new, and is not yet over.

Second, how effectively were the policies of the Nanking government carried out? All of China's twentieth-century governments tended to substitute legislation for actual reform and were usually too busy coping with crises to build responsive administrative structures. That is still very much the case. In the Nanking period announced policies were often actually initiated, and a few were even implemented, particularly if they applied mainly to the coastal and a few inland cities. But one cannot assume that a given policy was carried out. It is dangerous to make the leap from pronouncement to realization without evidence. What patterns does the evidence about scientific policy and its implementation reveal? What evidence do we have that policy was backed by substantial funds not of foreign origin? These are among the many questions that Ka-che Yip has raised in his commentary on this group of essays.

Third, why did young people decide to become scientists? We have had enough superficial pictures of scientists as puppets: uprooted intellects formed abroad, at the same time anointed agents of national progress and units of manpower in a massive effort to survive, a pool of inexpensive data collectors in the eyes of foreign colleagues. What did the formative parts of their careers look like from their own points of view?

James Reardon-Anderson has reminded us in an unpublished paper that in the 1920s the physical sciences were still generally considered philistine. In the 1930s elite families often discouraged their sons from going into fields that might not lead to civil service appointments.

An example of Tsui-hua Yang's suggests that not all decisions were based on values widespread in the society. Hu Shih and other May Fourth intellectuals had argued around 1920 that traditional historiography was inductive and data-oriented, like modern science. "Inductive" is an odd word to apply to the writings of early historians who tell us frankly that they are constructing exemplary accounts of good and bad characters for the moral guidance of readers. "Data-oriented" is a curious way to describe their use of sources, fastidious in some ways and promiscuous in others. But this claim was useful to reformers arguing that there was no absolute break between tradition and modernity. In 1941 we find Yang Chung-chien turning the simile inside out and recalling that he had found geological stratigraphy attractive because it was like traditional history. I doubt that anyone else ever accepted that particular wooden nickel. We will want to keep in mind that, in a disjointed time, many career decisions resulted from what now seem highly individual proclivities.

It is still not too late to learn from a fairly large group of survivors what attracted them while others ignored the sciences, and what kept some of them doing research in a culture that gave greater status to administrators. But soon it will be too late.

It is heartening to report that this work of preserving memories has begun. Journals such as *Chung-kuo k'o-chi shih-liao* frequently publish scientists' recollections of retired or deceased colleagues, and autobiographical memoirs by those inclined to write them. Such records are part of a public process that discourages frank recording of raw material, however, so that their greatest value is often as sources for the formation of modern legends. They are seldom disciplined by the probing and cross-checking that lies at the heart of oral history.

A few skilled historians have seen the importance of systematic interviewing. Laurence A. Schneider and William J. Haas have been fortunate enough to do substantial work in China with the help of colleagues there. Haas has begun work on a *Dictionary of Chinese Scientific Biography* based to a considerable extent on his notes. Then there is the remarkable project of Zhong Shaohua (Chung Shao-hua) of the Beijing Academy of Social Sciences. Zhong is harvesting the recollections of scientists, mostly between eighty-five and ninety-five years old, in repeated intensive interviews over a considerable period in a number of Chinese cities. The Research Institute for the History of Natural Science, Academia Sinica, has begun an oral history project that can draw upon substantial resources. But these are only minimal first steps. An adequate history of science and medicine in twentieth-century China cannot come from official documents alone. Living memories are essential.

Acknowledgements

The conference on "Research and Education in Twentieth-century China" was organized by my fellow editors, John Z. Bowers and J. William Hess. Support from the Rockefeller Archive Center (North Tarrytown, New York) made the conference possible and let us engage Barbara Congelosi to prepare the manuscript for publication. Pi-Chao Chen, Joseph W. Ernst, Arthur Kelman, Choh-ming Li, Michel Oksenberg, Leo A. Orleans, and James Reardon-Anderson were also present at the conference. Stanley Rudbarg made a stenotype record of the discussions, and we looked forward to including a summary of the transcription in this volume. His unfortunate death shortly after the conference made that impossible, but those who were at Pocantico Hills will remember his presence.

PART I

REPUBLICAN CHINA

PART I

REPUBLICAN CHINA

Genetics in Republican China

Laurence A. Schneider

> The most exciting publication I encountered at that period was a [1933] paper by Dobzhansky on geographic variation in lady-beetles. I exclaimed, "Here is finally a geneticist who understands us taxonomists!"
>
> *Ernst Mayr, "How I Became a Darwinian"*[1]

This paper describes and analyzes the development of genetics education and research in China from about 1920 to 1945. It transgresses the boundary of 1937 set by the conference panel because genetics, a relatively late-blooming field in China, showed significant signs of maturity during the wartime years. The body of the paper is comprised of a narrative of early and exemplary scientific careers and a reconstruction of the development of experimental genetics at Yenching University and of plant genetics and breeding at the University of Nanking. With no slight intended, at least a dozen well-trained, competent, and noteworthy geneticists will not be mentioned. Nor will all of the schools in which genetics was taught and researched enter into the discussion. My focus on Yenching and Nanking universities is a function of their unrivalled superiority during this era in terms of the educational backgrounds of

---

[1] From *The Evolutionary Synthesis,* ed. Mayr et al. (Cambridge, Mass., 1980), 419.

their faculty, the quality of the education they provided, and the quantity and quality of their published research.

For the reader's convenience, introductory fare and "conclusions" are supplied at the outset. My most general conclusion answers the question: How well did genetics education and research develop in China? My answer is: *qualitatively,* very well indeed, as evidenced in the training of students through the master's level, preparing them for the best doctoral programs abroad, and thereby creating researchers whose published work met the highest standards of the field. Qualitatively, very well, because the teaching and research of Chinese geneticists were clearly in the mainstream of genetics as it moved from the classical Morgan school to population genetics and the so-called evolutionary synthesis. *Quantitatively,* the geneticists—indeed all experimental biologists— constituted an elite minority vis-à-vis a preponderance of taxonomists in biology, and all the more so in relation to the number of physical scientists trained and working in China before 1949.

Using a topical approach, it will be helpful to summarize at this juncture the key points that enhance our understanding of how China achieved what it did in genetics and the character of that achievement.

The Central Role of the Foundations

In an earlier paper, I argued that during the Republican period the form and content of Chinese science, education, and research were shaped by the advice, the model institutions, and the abundant funds provided by the Rockefeller Foundation (mainly, its China Medical Board, or CMB) and the China Foundation for the Promotion of Education and Culture.[2] This paper shows that, particularly in the case of genetics, these two foundations were omnipresent and essential. Virtually every geneticist, throughout the course of graduate training and postdoctoral research, received a year or more of strategically important support from one or both of the foundations, for work either in China

---

[2] For a discussion of the role of these foundations and a fuller bibliography of relevant archival and published material, see my: "The Rockefeller Foundation, the China Foundation, and the Development of Modern Science in China," *Social Science and Medicine* 16 (1982): 1217–21; and "Using the Rockefeller Archives for Research on Modern Chinese Natural Science," *Chinese Science* 7 (1986): 25–31.

or in the United States. At Yenching, undergraduate and graduate biology and genetics were nurtured by the CMB because of their functional ties with the Rockefeller Foundation's Peking Union Medical College (PUMC). From 1925 on, most of the medical students enrolled by PUMC received their B.A. or M.A. from Yenching. And starting that same year, the Rockefeller Foundation and the China Foundation cooperated in the creation and maintenance of the University of Nanking's important exchange program with Cornell University.

It follows, then, that we should ask here whether the "Johns Hopkins model" played any role in the development of genetics. In my earlier paper, I argued that this model initially guided the CMB's ambitious plans to upgrade higher education in the natural sciences throughout China. Where medicine was concerned, this model of science was thought of as clinical practice wed to laboratory research, an activity devoted to the expansion of knowledge, to the pursuit of basic explanations of nature, and to the discovery of laws. In China, this model was generally taken to be a guide for the practice of "pure" as opposed to "applied" science. Insofar as Chinese development of genetics is concerned, I found no reference to this model as such. Rather, the field was developed in accordance with the standards exemplified by its American leaders (at Columbia and Cornell), following out its own internal logic. Furthermore, the Nanking genetics program was initially formed and later enhanced in the hope that its "applied" potential would lead to improved crops. It did. But the Cornell plant genetics program was broadly based and combined rigorous theoretical training with experimental techniques, so as to diminish if not erase conventional distinctions between pure and applied science.

Institutional Networks

Due largely to the initiative of the Rockefeller Foundation, genetics education and research occurred within two networks. In the first one, students and faculty moved between Soochow and Yenching universities. (Soochow was the base of N. G. Gee, an American biologist and the director of the CMB's influential science education program, who is discussed in depth in William Haas's paper in this volume.) From Yenching, students and faculty moved between T. H. Morgan's genetics programs at Columbia and, after 1928, the California Institute of Technology. The second network was a more formal one between the University of Nanking and Cornell.

Within these networks there were three noteworthy trends, characteristic of genetics as well as other areas of the life sciences. First, Chinese scientists supplanted Westerners in both Chinese and missionary schools. Second, Chinese scientists became collaborators, as opposed to passive students, of leading Western geneticists and, on returning to China, continued to communicate and sometimes collaborate with them. This provided an important outside source of community and direction for genetics in China. And third, Chinese geneticists established new programs and created research institutions where none had existed before. This was very much as the foundations had planned it. Whatever might be said about the "cultural imperialism" of American philanthropy, in China the foundations' central policy was to create an elite corps of doctoral degree-holders in the sciences who would liberate Chinese education from resident Western educators (generally a mediocre lot) and build a new set of state-of-the-art science programs.

The Dominance of Taxonomy

Although the development of systematic and taxonomic biology is more fully explored in another paper in this collection, there are several related points that should be noted here. With respect to professionals, experimental biology was numerically dominated by the fields of systematic and taxonomic biology, and this affected the scale of its development during the Republican period. Moreover, the development of experimental biology and genetics can only be fully understood in relationship to the naturalist mode.

Why, when Chinese life sciences were being planned so self-consciously, did naturalists come to outnumber experimental biologists (by at least five to one, if American Ph.D.-holders are counted), and geneticists in particular (by ten to one)?[3] Why did the planners of curricula, institutes, and overseas graduate work not feel compelled to put more of their resources and energies into that swift mainstream of Western biology characterized by the experimental, reductionist mode? Perhaps the answer lies in the notion of "local" or "particular" science, as opposed to "universal" science. This notion was brought to my attention by Tsui-hua Yang's work on geology for this panel and also appears in this collection in William Haas's essay. Chinese intellectuals of the

---

[3] Tung-li Yuan, comp., *A Guide to Doctoral Dissertations by Chinese Students in America 1905–1960* (Washington, D.C., 1961).

Republican period spoke fervently to this question, and, I am convinced, thereby influenced the science and education policies of both the foundations and the leaders of major Chinese institutions. In brief, these intellectuals argued that the priority of Western science in China should be the study of nature as it is manifest in China: China's geology, China's flora, China's fauna. The potential practical value of this approach is obvious. The nationalistic and nativistic sources of these ideas warrant further inquiry.

As far as the foundations' policies are concerned, there is no doubt about the uneven support given to naturalist biology. For example, both the Biological Research Laboratory of the Science Society and the Fan Memorial Laboratory were almost completely devoted to taxonomy. Both were the costly wards of the China Foundation. The CMB clearly fostered China's best geneticists; nevertheless, my own research has led me to conclude that the CMB, guided by Gee, purposely channeled resources disproportionately toward taxonomic curricula and research. In 1933, this policy was noted disapprovingly by W. E. Tisdale, one of the Rockefeller Foundation's program evaluators.[4] However, by that time a trend had been set and reinforced. China's earliest scientific entrepreneurs—Ting Wen-chiang 丁文江 , Ping Chih 秉志 , and Hu Hsien-su 胡先驌 —made their "naturalistic/localistic" biases felt first and most forcefully at the China Foundation and at key education and research institutions. Foreign science advisors such as J. M. Coulter from Chicago and J. G. Needham from Cornell were themselves from the naturalist tradition. Throughout the 1920s they admonished their Chinese colleagues to beware of "borrowed biology" and to focus on knowledge of nature in China. And when the Chinese Academy of Sciences (Academia Sinica) was opened in 1928, the inaugural speech of one of its directors, the historian Fu Ssu-nien 傅斯年 , supported this position under the banner of the nationalization of science. The question to be considered here is, did this emphasis on the naturalist tradition qualitatively affect the practice of experimental biology and genetics in China?[5]

---

[4] W. E. Tisdale, "Report of a Visit to Scientific Institutions in China" (unpublished paper, December 1933; Rockefeller Archive Center [hereafter, RAC]: 601.D/B 40), 5.

[5] J. M. Coulter is cited by J. W. Dyson in "The Science College," in *Soochow University*, ed. W. B. Nance (New York, 1956), 54–55. See

The naturalist and the experimental traditions of biology are not necessarily incompatible. Recently, for example, in a discussion of Barbara McClintock's genetic work, Stephen J. Gould argues that the most fundamental of biological issues—"the origin and development of organic form"—can only be addressed through a combination of naturalist and reductionist approaches.[6] And historians of science now conventionally speak of naturalist and genetic research traditions converging in an "evolutionary synthesis" during the period examined here. It is generally agreed that this synthesis, which has led to a substantially sounder understanding of the origin of species, was the *joint* product of, *inter alia*, the field naturalist and the Mendelian geneticist, the biometrician and the embryologist. In spite of this, it is also generally agreed that there was considerable resistance to or ignorance of this synthesis among the naturalists.[7] Were Chinese geneticists aware of and did they participate in the mainstream that led to this synthesis? Did the Chinese preoccupation with catch-up taxonomy impede them?

The geneticists offered as illustrations below each respectively entered the mainstream at a point nearer the synthesis than their predecessors. This is hindsight's way of saying that they trained and researched at Columbia under Morgan's influence and then at Cal Tech under Dobzhansky's: study of Morgan's classical genetics (Mendelism plus chromosome theory) came first, followed by work in cytogenetics (embryology plus classical genetics) and, finally, by population genetics (cytogenetics plus evolutionary theory). There is also evidence to support the contention that some of the Nanking geneticists, who studied at Cornell, were more likely than others in the field to have had the sort of broad training and insight that allowed them to move between their

---

also J. G. Needham, "Report to the China Foundation, June 1928," *China Foundation [Annual] Report*, no. 3 (1924), app. 1; and *Fu Ssu-nien hsüan-chi* [Selected works of Fu Ssu-nien], 10 vols. (Taipei, 1967), 3: 475–87 (the original is in the introduction to *Li-shih yü-yen yen-chiu-so chou-k'an* [Historical and philological research institute weekly] 1, no. 1 [October 1928]).

[6] "Triumph of a Naturalist," *New York Review of Books*, 29 March 1984, 6.

[7] See E. Mayr and W. B. Provine, eds., *The Evolutionary Synthesis* (Cambridge, Mass., 1980).

cytogenetical studies and the implications of those studies for evolutionary theory.

Thus, the bias toward naturalist biology in China seems not to have impeded the qualitative development of genetics. Furthermore, we may conclude that among Chinese geneticists there was that "feeling for the organism" which acted as a bridge between the reductionist and the naturalist proclivities. From pioneer geneticist Ch'en Chen's 陳楨 goldfish, through the rice, wheat, and maize of the Nanking school, to T'an Chia-chen's 談家楨 ladybird beetles, genetics in China showed particular respect for the whole organism, for its history and its environment. Where did this sensibility come from? Early on, it could have come right from Morgan at Columbia, for I am told that he had a lifelong devotion to marine taxonomy, and that, when all was said and done, he was not a true reductionist.[8] Perhaps it came from Alice Boring, Morgan's superb student who taught at Yenching. She eventually gave up genetics research for taxonomy. And as a final "perhaps," it may have derived from the influence of the Chinese taxonomists themselves. They were a powerful lot, and they had a good case.

A Pioneer: Ch'en Chen and His Goldfish

Ch'en Chen's career illustrates many of the main influences on the biological sciences in Republican China. Ch'en himself became a force in the field, shaping modern biological education and the place of genetics within it. His thinking on genetics and his research established early on a dialogue within the growing biological research community. He strongly emphasized an experimental approach to the life sciences and by that means tried to relate hereditary processes to the whole organism in its environment. Ch'en encouraged the proliferation of ideas about hereditary mechanisms in relation to variation, speciation, and evolution.

Ch'en Chen (1894–1957) was educated in his native Kiangsu, first at the China National Institute (1912–14), then at the University of Nanking (1914–18), where he earned his B.S. Upon graduation, he became a Tsinghua fellow, prepared for graduate study in the United

---

[8] Garland Allen, *Thomas Hunt Morgan, The Man and His Science* (Princeton, 1978).

States, and entered Columbia University's M.A. program in biology in the spring of 1920. Although he specified "genetics" as his major, what inspired this peculiar choice is a mystery. E. B. Wilson was Ch'en's advisor and primary instructor for the half dozen biology courses he took during 1920–21. Wilson was by this time a major figure in the development of classical genetics. In 1905, he had laid a cornerstone of chromosome theory (simultaneously with Nettie Stevens at Bryn Mawr) by demonstrating that sex was determined by a specific chromosome. T. H. Morgan had been at Columbia since 1905, and thus Ch'en was at the center of genetics in one of its most exciting periods.[9]

Ch'en completed his non-thesis master's degree in 1921. He thereupon returned to Nanking, taking a job at National Southeastern University (SEU) where a new biology department was flourishing under the direction of Ping Chih (Ph.D., Cornell, 1918, in zoology) and Hu Hsien-su (Sc.D., Harvard, 1925, in botany). Both of these scientists were throughout their careers promoters of large-scale taxonomic projects, as well as spokesmen on behalf of scientific education and research in general. Ch'en Chen, as China's first link with the Morgan school, was given the responsibility of teaching the first genetics course in China. He chose as texts Babcock and Claussen's *Genetics in Relation to Agriculture* (SEU's biology department was part of the agriculture college) and T. H. Morgan's *Physical Basis of Heredity*. Twelve students helped him launch the course.[10]

In the early 1920s, SEU was unrivalled in its science faculty and their promotion of research in their respective fields. (Notably, there

---

[9] Columbia University, Graduate Student Record Cards of Ch'en Chen. See also Ch'en's obituary in *Sheng-wu hsueh-pao* [Biology journal] 1 (1958): 64–65.

[10] N. Gist Gee, "Southeastern University, Nanking" (RAC: CMB/ser. II/B 83), 13–14. See also *Kuo-li Tung-nan ta-hsueh i-lan* [Catalogue of National Southeastern University] (Nanking, 1923), 62, 66; and E. B. Wilson, "Report on Science at SEU" (5-page appendix to John Grant's "Report on Welfare Activities and the Teaching of Hygiene at the National SEU," 21 November 1921; RAC: CMB/ser. II/B. 83). It should be noted that Ch'en Chen at this time authored his own biology textbook, and it became the most popular and influential Chinese language biology text until 1949. See *P'u-t'ung sheng-wu hsüeh* [General biology] (Shanghai: Commercial Press, 1924).

were no foreigners in SEU's biology department.) The faculty's efforts were supported by the good offices of the Science Society of China (a private organization of Chinese scientists) and the ample funds of the China Foundation. In 1922 their cooperation resulted in the Science Society Biological Research Laboratory, in Nanking. Ch'en Chen and Ping Chih were the first and only researchers to work in the new facility's zoological section during its first year. Thereafter, the laboratory became an active research arm of SEU biologists as well as a haven for visiting researchers from throughout China. In 1925, Ch'en's monograph on goldfish was the first title to be published in the laboratory's distinguished research series. By that time, however, it was clear that both SEU's biology department and the laboratory were beginning to devote themselves exclusively to taxonomics. This was a feature of biology generally throughout China, and it surely took much of the limited resources and personnel away from the development of experimental biology and genetics.[11]

SEU's biology department and the Biological Research Laboratory were heavily dependent on subsidies from the China Foundation for plant, equipment, and special research projects. In 1926, SEU and other major schools further profited from the foundation's efforts to develop science through a special Science Professorship program. Between 1926 and 1937, seventeen mature scientists were given generous salaries and research budgets for renewable one-year periods. Sometimes, the program brought a scientist to a campus or institute where the foundation thought his presence would be most influential. In other cases, the foundation simply took over the expenses of a scientist in his current location. Ch'en Chen was the first recipient of one of these professorships, which he held at SEU from 1926 to 1928.[12]

The next year, Ch'en began the last phase of his prewar career. He was brought to Peking to head and develop the young Tsinghua University's biology department. Under the auspices of the China Foundation and the Chinese central government in Nanking, Tsinghua began, after 1925, to make the transition from a college to a full-scale university. The Rockefeller Foundation, through the CMB, was inter-

---

[11] *The Science Society of China: Its History, Origin and Activities* (Shanghai, 1931), 19–20.

[12] See *China Foundation Report*, no. 1 (1925–26), and especially no. 2 (1926–27), "Zoology at Tungnan University."

ested in contributing to the biology department of this new school as well, and so Ch'en was in a uniquely creative position. Between 1928 and 1935, he increased the department's faculty from six to thirteen, and he had at least three American Ph.D.s among them (from Chicago, Yale, and Cornell).[13]

Ch'en's teaching role as well as the subjects of genetics and experimental biology were prominent in the curriculum requirements for biology majors. He taught the required courses in cytology, experimental genetics, and genetics and evolutionary theory. The latter course is one of the few in Tsinghua's catalogue that lists its texts (Sinnott and Dunn's *Principles of Genetics,* W. B. Scott's *The Theory of Evolution*) and its main topics: "the variability of living things; Mendel's laws of inheritance and their generalization; the material basis of heredity; the determination of sex; evidence for the theory of evolution (mutation theory; theory of use and disuse [i.e., Lamarckism]; natural selection)." These were not courses designed for an agriculture school.[14]

Ch'en's happy situation at Tsinghua was made complete in 1933 with the opening of a massive new biology building, paid for by the CMB and a matching (if reluctant) grant from the central government. In 1933, W. E. Tisdale's evaluation of Ch'en's department was as unequivocal as Tisdale ever got: "As far as potential possibilities exist, this department would seem to be the strongest I have seen yet in China."[15]

Ch'en Chen's entrepreneurial renown was matched by his informal

---

[13] See the following: *Ch'ing-hua i-lan* [Tsinghua catalogue] (Peking, 1927); *Pei-p'ing ko ta-hsüeh te chuang-hsiung* [An outline of universities in Peking] (Peking, 1929–30); *Kuo-li Ch'ing-hua ta-hsüeh erh-shih chou nien chi-nien k'an* [Twentieth anniversary volume for National Tsinghua University] (Peking, 1931); *Kuo-li Ch'ing-hua ta-hsüeh* [National Tsinghua University] (Peking, 1935); and *Ch'ing-hua ta-hsüeh hsiao shih kao* [Draft history of Tsinghua University] (Peking, 1981), 205–9. See also the Rockefeller Foundation, *Annual Report* (New York, 1929), 226; and N. G. Gee, letter to Tsinghua president Tsao, 14 February 1926 (attachment to Roger Greene, letter to M. K. Eggleston, 26 March 1926; RAC: CMB/ser. II/B. 85).

[14] "Sheng-wu hsi" [Biology department], in *Kuo-li Ch'ing-hua ta-hsüeh i-lan* [National Tsinghua University catalogue] (Peking, 1935).

[15] Tisdale, "Scientific Institutions in China," 27.

fame as the man who used goldfish as his experimental animal. Though he published little, he did provide us sufficient insights into his thoughtful, careful work using *Carrasius auratus,* the common goldfish. Ch'en began to use this fish in 1923 when he was outlining plans for a long-term project meant to explore issues of heredity, variation, and speciation. He said he chose this animal because it would permit him to synthesize experimental approaches to embryology and breeding. Ch'en was quite aware how important a carefully selected experimental organism could be in these areas of investigation; witness the place of the fruit fly (*Drosophila melanogaster*) in chromosome theory. If the goldfish bred much slower and required more care than the fruit fly, it had its special advantages, too: Ch'en compared the goldfish to Darwin's pigeons and argued that the fish were even more valuable for demonstrating the concept of variation. Since the Sung dynasty, China's passion for goldfish breeding had led to a degree of variation equal to that among the pigeons of the world; "but the goldfishes' origins in the natural state are far easier to determine and the historical material regarding its varietal formations is even more complete."[16]

Using the rich textual sources available, Ch'en was able to study the timing and locale of goldfish variations; using Mendelian-Morgan genetics, he tried to note the occurrence of mutations. During 1924–37, using goldfish he had bred or whose lineage he knew, he performed a series of crossbreeding experiments. In 1928, the results of one series of experiments were reported in *Genetics,* the major forum of classical genetics in America. And in 1934, another report was published in J. B. S. Haldane's *Journal of Genetics,* in England. In these pieces, and in less technical ones he published in Chinese-language journals, Ch'en showed a breadth of concern that perhaps outstripped his training. However, his most technical pieces are fine applications of standard ways to establish the relationship between phenotype (characters such as color or shape of an organ) and genotype (an organism's genetic inventory, which may affect the appearance of a character). Convinced that muta-

---

[16] Ch'en Chen, "A History of the Domestication and the Factors of the Varietal Formation of the Common Goldfish, *Carassius auratus,*" *Acta Zoologica Sinica* 6, no. 2 (1954): 89–116 (English translation in *Scientia Sinica* 52 [June 1956]: 287–321); see also his "Chih-yü te pien-i yü t'ien-yen" [Variation and evolution of the goldfish], *K'o-hsüeh* [Science] 10, no. 3 (1925): 304–30.

tion was the key to variation and speciation, in the manner of de Vries and Morgan, Ch'en was especially interested in exploring the expression of mutation. In his *Genetics* article, for example, he addressed the issues of somatic mutation and mosaic mutation (wherein mutant genes occur in only part of the organism's body, resulting in the appearance of one part as normal, the other as mutant).[17]

One of the unusual features of Ch'en's otherwise conventional inheritance studies is that they go beyond the immediate laboratory experiment to ask historical questions. With his knowledge of the history of breeding techniques and of varieties of goldfish, Ch'en tried to explain, for example, why a variety of fish was rare when, according to Mendelian principles and natural selection, it should be quite common. As for Ch'en's own work in experimental embryology, since he left virtually no published record of it there is no way to make an assessment.

It was important to Ch'en Chen, and, I am sure, to his delighted Chinese audience, that his experimental animal was "Chinese." Indeed, in his study of the evolution of the goldfish, he concluded that it most certainly originated in China and not, as some claimed, in Japan or Europe.

Genetics at Yenching University

Yenching's rapid development of excellence began in 1925, with the implementation of the CMB's earlier decision to turn PUMC's preparatory department over to Yenching. That decision was part of a larger and more ambitious one, to enhance "premedical" science education in major schools throughout China. Between 1923 and 1937, Yenching's biology department grew from a staff of two undistinguished foreign teachers to a superbly trained, versatile, and actively researching team of ten, which included, at its peak, four doctors of biology trained in America. Since the bulk of PUMC students came from Yenching, the latter was nurtured with large endowments and grants. Under CMB

---

[17] Ch'en Chen, "Transparency and Mottling, a Case of Mendelian Inheritance in the Goldfish *Carassius auratus*," *Genetics* 13 (1928): 434–52; and "The Inheritance of Blue and Brown Colours in the Goldfish *Carassius auratus*," *Journal of Genetics* 29 (1934): 61–74.

influence, a link also developed between Yenching and Soochow University.

From 1926 on, Yenching's biology faculty were all Chinese, with the important exception of Alice M. Boring (1883–1955). Boring was a preeminently qualified scientist whose curriculum vita reads like a history of early twentieth-century mainstream biology. Between 1900 and 1910, she received her B.A., M.A., and Ph.D. in biology from Bryn Mawr, working under T. H. Morgan and Nettie M. Stevens. Her dissertation and collaborative work with her mentors ranged from experimental embryology to chromosomal genetics. She had the distinction (rare for any biologist, and especially for a woman) of a year's research fellowship at the Naples Zoological Station, a world-renowned center. Her teaching career began with a one-year appointment at Vassar in 1907–8, which was followed by ten years on the faculty at the University of Maine (1908–18). In Maine, she collaborated with Raymond Pearl at the Maine Experimental Station on some of his pathbreaking work in biometry and population genetics. When Boring accepted a teaching position with PUMC's premedical science program, in 1918, she had already published fifteen papers, including her disseration. She taught there for two years, then returned home to teach at Wellesley for another three; she went back to Yenching in 1923 as a full professor and head of the biology department.[18]

For the next two years, the department continued to serve the preparatory program for PUMC. Boring's most important colleague at this time was Ch'en Tze-ying 陳子英 (b. 1897), who had received his B.S. from Soochow University in 1921 and, doubtless due to the influence of N. G. Gee, was subsequently brought into the rigorous teaching program at PUMC/Yenching. Ch'en taught there for four years, simultaneously working under Boring to earn his master's, the first awarded by Yenching in biology. His thesis project was a chromosomal study of two kinds of fruit fly mutations. Supported by a Rockefeller Foundation

---

[18] Alice M. Boring, "Publications of Alice M. Boring" (1-page typescript; Bryn Mawr Archives); Dwight Edwards, *Yenching University* (New York, 1959), 160; "A Brief History of the Department of Biology of Yenching University" (34-page pamphlet, Yenching, December 1931; RAC: RG.1/ser. 601/B. 41); *American Men of Science* (5th ed., New York, 1933), 14; and N. G. Gee, letter to R. M. Pearce, 5 February 1929, section on "Yenching University" (RAC: RG.1/ser. 601/B. 40).

fellowship, Ch'en was then sent to Columbia, where he received an M.A. in biology in 1926. In 1929, under T. H. Morgan's supervision, he completed a doctoral dissertation that, fifteen years later, was still being referred to as a classic by scientists such as Joseph Needham.[19] Ch'en's research, on "The Development of Imaginal Buds in Normal and Mutant *Drosophila melanogaster*," illustrates a major emphasis of Morgan's school at that time. It was cited regularly throughout the 1930s by other geneticists, including young Chinese geneticists, such as Li Ju-ch'i 李汝祺 (b. 1896), who were to make their mark on Yenching.

An imaginal bud or disc is an inchoate organ, visible on a developing embryo or pupa. Ch'en's research was an exemplary study of how genes affect the development of characters at the earliest stages of embryonic growth. Genetic researchers were no longer satisfied with the generalization that a chromosome, gene, or set of genes transmits a visible character. Now they wanted to know when and where in the developmental process the effects of those shaping forces were first manifest. Ch'en studied the expression of certain mutant genes and found that the effects for their related character expressions "could be observed in the very earliest condition of the imaginal buds." Before Ch'en's work, when an organ of the fruit fly was, for example, found to be reduced or absent, "it was not known whether its imaginal disc showed a corresponding reduction or whether it was full size in the larvae and pupae, and failed to carry through to the later stages." With his precise studies of mutant gene expressions, Ch'en made a significant contribution to the interpretation of complex expressions such as mosaics, intersexes (a combination of male and female characters), and gynandromorphs (half the body showing one set of characters, the other half exhibiting another).[20]

Ch'en Tze-ying returned to a mature Yenching biology department as an associate professor in 1928–29. The China Foundation supported

---

[19] Joseph Needham and Dorothy Needham, eds., *Science Outpost: Papers of the Sino-British Science Co-operation Office 1942–1946* (London, 1948), 226; and Columbia University, Gradute Student Record Cards of Ch'en Tze-ying. (This romanization of Ch'en's name is used throughout his records and publications in America.)

[20] Ch'en Tze-ying, "On the Development of Imaginal Buds in Normal and Mutant *Drosophila melanogaster*," *Journal of Morphology and Physiology* 47 (March-June 1929): 135–200.

him and his research for two more years, after which he became head of the biology department at National Amoy University, where he was charged with the task of developing it into a department of Yenching's quality but with an emphasis on marine biology. I have no knowledge of any published research by Ch'en from this point. However, in 1945, when Joseph Needham met Ch'en in Amoy, he noted that Ch'en had the "best *Drosophila* stocks in China,"[21] which I read as an indication of Ch'en's continuing research efforts.

Upon Ch'en Tze-ying's return to Yenching in 1928, he found that Wu Chin-fu 吳經甫 (b. 1896) had joined the biology faculty and become its chairman. Wu was another Soochow alumnus. An entomologist, he earned both B.A. and M.A. under N. G. Gee (1914–18), and then a doctorate from Cornell in 1922. Wu returned to Soochow in 1923 as head of its biology department and director of the Biological Supply Service, which Gee had founded and fostered for many years. Clearly, placement of faculty was one means by which Gee was able to shape biology at Yenching. He also exerted his influence through the handpicked students who were sent to Yenching from Soochow, as well as through his collaborative work with Yenching faculty and through the economic power he wielded as the CMB's science education advisor.

Gee and his Soochow colleague J. Y. Dyson were proud of the role their biology graduates played in deparments around the country: during the late 1920s, fifteen of them were either the heads or ranking associates of their departments. With the exception of two, however, all taught at missionary schools. (Of the fifteen, at least seven obtained American doctorates, three of those from Cornell). Dyson thought it equally significant that, over the 1920s and 1930s, eight of the Soochow biology faculty had received Rockefeller scholarships, mainly for doctoral training at American institutions. He was quite explicit about the intended developmental role of these scholarshps: they were "the foundation upon which was to be built the structure of the future staff of Doctors of Philosophy to man institutions and to head up their science departments."[22]

Clearly, the major concern of biology at Soochow was taxonomic, and this reflects Gee's personal devotion to freshwater biology. Dyson reports that a formal policy to develop in that direction, rather than

---

[21] Needham and Needham, *Science Outpost*, 226.
[22] Dyson, "The Science College," 52–53.

toward experimentalism, was established when Gee began implementing the CMB's nationwide science education plan. Dyson notes that around 1923, John M. Coulter, former head of Chicago's botany department, was engaged as a consultant to Soochow's science program. He advised "more serious attention to taxonomy and studies of the local flora, as such data were basic to the botanical sciences in any country."[23] Soochow responded successfully with specialized training of its faculty and students. Coulter's advice probably also influenced Gee's science education policy throughout China.

Soochow's close connection with Yenching is further demonstrated by the flow of students between them. From 1923 to 1933, Yenching's biology department awarded twenty-two master's degrees (half during 1923–30, and the other half during 1931–33), with nine of them going to Soochow graduates. (Another seven successful master's students came from Yenching's undergraduate program, and the rest from various missionary schools; none was from a Chinese school.) Of the eleven Yenching students awarded master's degrees in biology during 1923–30, at least three went on for doctorates; and of the eleven during 1931–33, at least five went on for doctorates. Four of these eight Ph.D.-holders had done their undergraduate work at Soochow. (In 1929, Soochow had a total student population of 708; Yenching had 743.) This impressive record was fostered by funding from the two foundations: during the 1923–33 period, the Rockefeller Foundation supported thirteen of Yenching's twenty-two biology gradute students with one- or two-year fellowships, while the China Foundation gave advanced-study or research fellowships to another seven.[24]

In 1927, Li Ju-ch'i, another Morgan-school geneticist, was added to the Yenching biology faculty. From that point on, he was responsible for both genetics classes and theses. Alice Boring, from 1926 onward, turned to taxonomy, as evidenced in a series of collaborative publications on Chinese fauna with N. G. Gee, Wu Chin-fu, and others. Boring's reorientation may account to some degree for the paucity of geneticists coming out of Yenching's program. In spite of Yenching's prowess in genetics, between 1923 and 1933 only four of the twenty-two

---

[23] Ibid., 54–55.
[24] Ibid., 50, 53, 56; and "Brief History of the Department of Biology of Yenching University," *passim*.

graduates produced master's theses in genetics. Of the eight who earned doctorates, two specialized in genetics.[25]

Li Ju-ch'i's presence on the biology faculty marks the beginning of the department's most mature and productive period. He brought remarkable energy to the task and, like his colleagues, a superb education and pedagogy. He attended Tsinghua College between 1911 and 1919, then went to Purdue's agriculture school, where he underwent a kind of conversion from agronomy to genetics. His B.S. thesis was a routine piece of work on the pathology of hog cholera. But his 1924 M.S., also from Purdue, was a doughty piece of *Drosophila* genetics. The thesis brashly corrected some of T. H. Morgan's work.[26]

Morgan was apparently impressed enough by Li Ju-ch'i to accept him as both an M.A. and then a Ph.D. student, from 1923 to 1927. Li's doctoral research (published in *Genetics* in 1927) was closely akin to that of his Columbia fellow-student, Ch'en Tze-ying, whom Li acknowledges. Again, the general research goal was to relate genotype and phenotype through the mediation of embryonic development. The specific goals of Li's work grew out of the work of C. B. Bridges, one of Morgan's key students and co-workers at Columbia. Bridges had Li explore the effects of various kinds of gene deficiencies, especially on live birth and longevity. Like Ch'en Tze-ying, Li paid attention to the chronology of embryonic development vis-à-vis the development of imaginal discs. In addition, Li varied environmental factors (e.g., food, temperature) and genetic deficiencies (by eliminating sections of a chromosome or a whole chromosome). How these sets of factors interacted revealed in greater and more accurate detail both the degree of

---

[25] "Brief History of the Department of Biology of Yenching University," 10–19; and Yuan, *Guide to Doctoral Dissertations.*

[26] Columbia University, Graduate Student Record Cards of Li Ju-ch'i; see also Li Ju-ch'i, "The Pathology of Hog Cholera" (B.S. thesis, Purdue University, 1923), and "A Study of the Vermilion Stock of *Drosophila m.* with Special Reference to the Recurrence of Balloon Wings and the Appearance of 'Matted Bristles'" (M.S. thesis, Purdue University, 1924). Since I completed the original research for this essay and submitted it for publication in 1984, I have been able to do additional research in the People's Republic, including interviews with key scientists. The material on Li Ju-ch'i in this section was corroborated in my interview with Li in Beijing in August, 1984.

independence of developmental processes and how they depend on different genetic and environmental factors.[27]

After returning to China, Li Ju-ch'i's research was in part devoted to a collaborative effort on linkage experiments with E. B. Bridges, who died in 1938. ("Linkage" is the association of certain genes, owing to their location on the same chromosome.) The work Li began in his dissertation research was further developed with graduate students at Yenching, under the sponsorship of the China Foundation.[28]

The Genetics of T'an Chia-chen

T'an Chia-chen (b. 1910) was Li Ju-ch'i's most accomplished student and China's most accomplished geneticist of this period. Under Li's prescient supervision, T'an was prepared to take full advantage of the training he received at Cal Tech, one of the major research centers contributing to the "evolutionary synthesis" and the development of population genetics. Through T'an, Chinese genetics continued to remain abreast of the major currents in genetics by direct contact and collaboration with its principal exponents.

T'an received his B.S. from Soochow in 1930, and for the next two years, sponsored by a Rockefeller fellowship, he studied for his M.A. at Yenching. Li introduced T'an to classical genetics via breeding and inheritance-pattern experiments using ladybird beetles (*Harmonia axyridis*). Based on this work, T'an and Li collaborated on a series of papers which became the foundation of T'an's later contributions to population genetics. During 1932–33, T'an was a Rockefeller research fellow at Soochow, where he taught part-time and continued his impressive research on the wing color patterns of his ladybirds. Tisdale's

---

[27] Li Ju-ch'i, "The Effect of Chromosome Aberration on the Development in *Drosophila m.*" (Ph.D. diss., Columbia University, 1927), *Genetics* 12 (May 1927): 1–59.

[28] Li Ju-ch'i and Tsui Yu-lin, "The Development of Vestigial Wings Under High Temperature in *Drosophila m.*," *Genetics* 21 (May 1936): 248-63; Li Ju-ch'i and E. B. Bridges, "Genetical and Cytological Studies of a Deficiency (Notopleural) in the Second Chromosome of *Drosophila m.*," *Genetics* 21 (November 1936): 788–95, and 23 (November 1938): 110.

"Report" recognized the importance of T'an's work and also noted that T'an had developed a series of exhibits showing inheritance factors, arranging them in small boxes that were distributed by the Biological Supply Service to teachers throughout China.[29]

From 1933-34 to 1939, T'an studied at Cal Tech, continuing under the sponsorship of the Rockefeller Foundation. Under the direction of Th. Dobzhansky, he completed his doctoral dissertation in 1936 and then continued his research there in collaboration with his mentor as well as with A. H. Sturtevant (one of the original Morgan group) and others. Cal Tech's Kerckhoff Laboratories, where T'an worked, was subsidized by the Rockefeller Foundation.

T'an's training at Cal Tech, as reflected in his published papers, was a step-by-step preparation for his attack on questions concerning the genetic character of "race" (a "race" being a genetically distinct population that is found in a specific territory, but in which individuals have varying Mendelian genotypes), the mechanics of speciation, and, hence, the genetic basis of natural selection and evolution.

T'an's first work focused on the genetic differentiation of varieties of *Drosophila*. There were two techniques for doing this, and T'an mastered both. First, there was direct cytological observation of chromosomes and description of their morphology. This was a technique made possible by the discovery in the late 1920s of the giant salivary gland chromosomes of the fruit fly. Second, there was the complementary method of chromosome mapping, pioneered by Sturtevant twenty-five years earlier. Mapping is based on a fundamental tenet of classical genetics, namely, that genes are arranged linearly on a chromosome. By making statistical observations of the appearance of specific characters, such as color or wing shape, inferences could be drawn about the relative location of a particular gene on a particular chromosome. When these two techniques are followed, as they were in T'an's dissertation, it is possible, for example, to differentiate genetically some races of

---

[29] Tisdale, "Scientific Institutions in China," 56; Li Ju-ch'i and T'an Chia-chen, "Variations in Color Patterns in Lady-bird Beetle, *Ptychanatis axyridis Pall*," *Peking Natural History Bulletin* 7 (1932): 175-93; and Li and T'an, "Inheritance of the Elytral Color Patterns of Lady-bird Beetle, *Harmonia axyridis*," *American Naturalist* 68 (1934): 252-65. Information about T'an Chia-chen, in this section and below, was corroborated and extended by my interview with T'an in Shanghai in August 1984.

Drosophila whose outer appearances are indistinguishable. In tandem, these sensitive techniques made it possible for T'an to conclude, among other things, that "racial differentiation partook [in the cases under study] of both sources of evolutionary variability—gene variability and variability in the gross structure of chromosomes."[30]

When T'an Chia-chen returned to China, he became a member of the rapidly growing science faculty of Chekiang University. The war had already driven the university to its southwestern wartime refuge in Meitan, Kweichow. Undaunted, perhaps even exhilarated by the new natural environment he found there, T'an eventually gathered under him a team of four senior researchers and seven assistants.

Funded in part by the Rockefeller Foundation and the Chinese Ministry of Agriculture, T'an directed a number of projects that centered on some fundamental issues in population genetics. Synthesizing his earlier work on the ladybird beetle with his work under Dobzhansky, T'an first isolated and characterized a race of the beetle, then began to account genetically for the specific, geographically defined populations of this race. The extraordinary color patterns of the beetles and the role of "mosaic dominance" in the inheritance of these color patterns made the study feasible and elegant.[31] Dobzhansky later canonized T'an's

---

[30] T'an Chia-chen, "Kuan-yü i-ch'uan te wu-chih chi ch'u wen-t'i" [On the question of the material basis of heredity], *Sheng-wu hsüeh t'ung-pao* [Biology journal], no. 1 (1957), 9; "Salivary Gland Chromosomes in the Two Races of *D. pseudoobscura*," *Genetics* 20 (July 1935): 401; "Identification of Salivary Gland Chromosomes in *D. pseudoobscura*," *Proceedings of the National Academy of Science* (Washington, D.C.) 21 (1935): 200–202; "Genetic and Cytological Maps of the Autosomes in *D. pseudoobscura*" (Ph.D. diss., California Institute of Technology, 1936); "Genetic Maps of the Autosomes in *D. pseudoobscura*," *Genetics* 21 (November 1936): 796–807; Th. Dobzhansky and C. C. T'an, "Studies on Hybrid Sterility. III. A Comparison of the Gene Arrangement in Two Species of *D. pseudoobscura* and *D. miranda*," *Z.I.A.V.* 72 (1936): 88–114; and A. H. Sturtevant and C. C. T'an, "The Comparative Genetics of *D. pseudoobscura* and *D. melanogaster*," *Journal of Genetics* 34 (1937): 415–32.

[31] T'an Chia-chen, "Inheritance of the Elytral Color Patterns of *Harmonia axyridis* and a New Phenomenon of Mendelian Dominance," *Chinese Journal of Experimental Biology* 2 (1942); "A Report on the Work of the Laboratory of Genetics and Cytology,

study by citing it in his classic book on genetics and evolution. In particular, he noted T'an's paradigmatic illustration of racial variation, which is a function of how often the alternate forms of just one gene combined with one another.[32]

T'an received a Rockefeller Foundation grant to complete and write up his ladybird studies at Columbia during 1945–46. At both Columbia and the Biological Laboratory at Cold Spring Harbor, he also did research on a related but unstudied question, namely, the genetic basis of sexual isolation (the dominance of intra- over inter-species mating). The results of his work in America were immediately published in *Genetics*.[33] T'an returned to China in 1946, just two years before the Lysenkoist chill descended and silenced him for the better part of the next decade.

The University of Nanking and Plant Genetics

From 1925 through the war years, the distinguished Agriculture College of the University of Nanking had the largest, most active teams of geneticists in China. It was in 1925 that the school's close, productive relationship with Cornell University began, with Cornell agreeing to send one visiting professor in plant breeding each year for five years and to provide opportunities for graduate study and research at Cornell. Cornell was a major center for genetics, starting with R. A. Emerson's

---

Biological Institute, National University of Chekiang, Meitan, Kweichow, China, 1942–44," *Acta Brevia Sinensia* 8 (December 1944): 5–21.

[32] Th. Dobzhansky, *Genetics of the Evolutionary Process* (New York, 1970), 270. This book is a revised edition of Dobzhansky's *Genetics and the Origin of Species* (New York, 1937), which contains numerous citations of T'an's research. Also see high praise for T'an's work and the assignment to him of a key role in the development of population genetics in R. C. Lewontin et al. (eds.), *Dobzhansky's Genetics of Natural Populations* I-XLIII (New York, 1981), 37–40.

[33] T'an Chia-chen, "Mosaic Dominance in the Inheritance of Color Patterns in the Lady-bird Beetle, *Harmonia axyridis*," *Genetics* 31 (March 1946): 195–210; "Genetics of Sexual Isolation Between *D. pseudoobscura* and *D. persimilis*," *Genetics* 31 (November 1946): 558–73.

work at the beginning of the century and continuing with the research of such outstanding scientists as Beadle and McClintock. The Cornell-Nanking program was supported by the Rockefeller Foundation International Education Board.[34]

The University of Nanking was able to take full advantage of the Cornell connection through the good offices of the American Committee for China Famine Relief, which in 1925 endowed the school with the immense sum of $700,000 in gold, followed by an additional contribution of $600,000 in 1932. These grants were used to enlarge the programs in plant breeding, plant pathology, and forestry. In 1920, the University of Nanking had fewer than 100 students. By 1928 it had more than 500, of whom 175 were in the Agriculture College. The university grew steadily, even during the war, and boasted a total enrollment of 1,100 in 1949.[35]

Cornell's direct influence on plant genetics at Nanking continued right up to the school's evacuation to Chengtu in 1937. At that point, Cornell professor H. H. Love had been head of the Nanking plant genetics department on two separate occasions for a total of four years. Funds for the visiting faculty program had been extended beyond 1930 by the Rockefeller Foundation; from 1934–35 onward, these funds were enhanced by the China Foundation and by the British Boxer Indemnity fund. This philanthropic interest in agricultural sciences was a result of concerted efforts to aid the Chinese central government's formulation

---

[34] W. P. Fenn, *Christian Higher Education in Changing China, 1880–1950* (Grand Rapids, Mich., 1976), 92–93; Raymond B. Fosdick, *The Story of the Rockefeller Foundation* (New York, 1952), 183–84; Chunjen C. Chen [sic] (Tsinghua College), "The History and Present Status of Agricultural Education in China" (25 October 1925; RAC: CMB/ser. II/B. 85),3; and H. H. Love and John Reisner, *The Cornell-Nanking Story* (Ithaca, New York, 1964). My understanding of the material in this entire section on the University of Nanking and plant genetics was greatly enhanced through correspondence with Professor C. C. Li and an interview with him in Pittsburgh in January 1985. Professor Li (Ph.D., Cornell, 1940), a plant geneticist and biometrician, was trained in the Nanking-Cornell program. During 1943–46, he was professor of genetics and biometry in the Agriculture College of the University of Nanking.

[35] Chunjen Chen, "Agricultural Education in China"; Fenn, *Christian Higher Education,* 92, 239.

of a policy toward the virtually ignored rural-agricultural sector. Embarrassed by the Communists' "rural reform" successes in Kiangsi as well as by government failures in dealing with natural disasters in the hinterland, the central government finally began to act after its successful anti-Communist campaigns in 1934. The Rockefeller Foundation's special "China Program" and the sudden rural orientation of the other foundations are linked with this political turning point.[36]

By the early 1930s, the Nanking plant-breeding and genetics program had gained a strong reputation for its crop improvement research, which had produced studies of rice diseases as well as genetic studies that resulted in the development of three varieties of high-yielding wheat more resistant to endemic diseases and parasites. In addition, the program now had growing numbers of significant Chinese faculty, such as Li Hsien-wen 李先聞 (b. 1902) and Wang Shou 王綬.[37]

Wang Shou's 1936 textbook, Chung-kuo tso-wu yü-chung hsüeh 中国作物育種學 [Crop breeding in China], gives some insight into genetics education at Nanking. An introductory text, it nonetheless provides quite a sophisticated presentation—in Chinese—of the fundamentals of classical genetics. It pointedly takes what E. Mayr calls the "hard heredity" line, i.e., that acquired characters are not inheritable. Notably, in addition to building his text on Western expertise (R. G. Wiggens, H. H. Love, Babcock, Claussen), Wang also cites the recent work of Chinese plant geneticists (Li Hsien-wen, Hao Ch'in-ming, Chao Lien-fang 趙連芳). The text is based on Wang's classroom lectures at Nanking before the war.

Equally impressive and insightful is the Chinese-language textbook I-ch'uan hsüeh 遺傳學 [Genetics], written by Wang's colleague Fan Chien-chung 范謙衷 in 1942.[38] Based on Fan's classroom lectures, this book takes a more historical approach to genetics. With careful atten-

---

[36] Schneider, "Development of Modern Science in China," 1220; Tisdale, "Scientific Institutions in China," 11; *China Foundation Report*, no.10 (1934–35), 38; British Boxer Indemnity *Report* (London, 1934–35 and subsequent issues); Rockefeller Foundation, *Annual Report* (New York, 1935), 321; Needham and Needham, *Science Outpost*, 115.

[37] *China Foundation Report*, no. 10 (1934–35).

[38] Chinese Ministry of Culture, Education, and Industry, *I-ch'uan hsüeh* [Genetics] (Chengtu, 1942).

tion to terminology (English equivalents given in parentheses), classical genetics from Weismann through Mendel and Morgan are presented in a clear, orthodox fashion. Going beyond classical genetics, this book introduces recent cytological research, biometrics, and radiation studies stemming from Muller's work.

Books such as these offer evidence that the genetics education of the "applied" plant geneticist was no less basic and thorough than, for example, that of the Yenching geneticists. (It is necessary to point this out because of the later tendency in the People's Republic to denigrate the role of theory—especially classical genetics—in plant breeding and to emphasize popular experience and trial-and-error technique.)[39] This is quite evident during the most developed stage of the Nanking genetics school, from 1937 to 1949, which was at that time able to draw on the skills of a dozen Chinese holding doctorates in plant genetics, most of them recently returned from Cornell, where they had been sponsored by China Foundation fellowships.

As the central government of China began to initiate policies toward agriculture, it drew on the plant-breeding expertise of the University of Nanking. Evidence of this is seen in the excellent journal *Nung-pao* 農報 [Farming], published by the Ministry of Agriculture and Forestry. This journal, which from 1935 on became a forum for the Nanking geneticists, details cooperative research projects, sponsored in part by the government, as well as extension work. In 1939, the Nanking plant-breeding program, now relocated in Chengtu, provided most of the staff of the Szechwan Agriculture Improvement Institute (an agency sponsored by the central government). The prolific work of the institute and the Nanking geneticists is documented in *K'o-hsüeh nung-yeh* 科學農業 [Scientific agriculture], first published by the Ministry of Agriculture and Forestry in 1943.

The Szechwan institute assembled a critical mass of geneticists headed by Li Hsien-wen (Ph.D, Cornell, 1930), a maize specialist. With the exception of J. L. Buck, from Cornell, and a few visiting foreign experts, both the institute and the University of Nanking's plant-breeding department were now completely staffed with Chinese, among them Chao Lien-fang (Ph.D., Wisconsin, 1927), a rice specialist; Feng

---

[39] For example, see Jack R. Harlan, "Plant Breeding and Genetics," in *Science in Contemporary China*, ed. Leo Orleans (Stanford, 1980), 295–312.

Che-fang 馮澤芳 (Ph.D., Cornell, 1935), a cotton specialist; and Ling Li 凌立 (Ph.D., Minnesota, 1937), a specialist in physiology and parasitology.

Detailed reports of the institute's work show it to have been devoted to a variety of discrete projects on theoretical questions in addition to long-term projects on specific food-plant problems. For example, Li Hsien-wen directed decade-long genetic studies of debilitating abnormalities such as "dwarfism" in local wheat crops and "polyhusk" in rice.[40] Other geneticists studied problems of interspecific hybridization, hybrid vigor, and male sterility. There was also work on induced polyploidy (using colchicine) and studies in haploid breeding.[41] (Haploids have only one set of unpaired chromosomes in each nucleus. Their potential advantage lies in the creation of immediate genetic uniformity, uniform lines, and easily identified genotypes. Haploid breeding has been fashionable in the People's Republic at least since 1966 because of its efficiency. I have seen no evidence that the fashion had already begun before 1949 among the Nanking geneticists.)[42] The sources cited

---

[40] See the special agricultural science issue of *Chin-ling hsüeh-pao* [University of Nanking journal] 11, no. 3 (1941); Pao Wen-k'uei et al., "Studies on the Inheritance of Dwarfness in Common Wheat" (Chinese with English summary), *K'o-hsüeh nung-yeh* [Scientific agriculture] 1, no. 1 (September 1943): 1–12; and Kuang Hsiang-huan et al., "Genetical Studies on the Polyhusks in Cultivated Rice (*Oryza sativa L.*)," *K'o-hsüeh nung-yeh* 1, no. 2 (December 1943): 119–24. (The English titles are used in the the original publication.)

[41] Li Hsien-wen et al., "Interspecific Crosses in *Setaria*: II. Cytological Studies of Interspecific Hybrids," *Journal of Heredity* 33, no. 10 (October 1942): 351–55; Pao Wen-k'uei et al., "The Inheritance of Pentaploid Wheat Hybrids, A Critique," *K'o-hsüeh nung-yeh* 1, no. 1 (September 1943): 23–24; Chin Tzu-chung, "The Inheritance of Some Quantitative Characters in the Interspecific Crosses of Wheat," *K'o-hsüeh nung-yeh* 1, no. 3 (March 1944): 204–18; Li Hsien-wen et al., "Cytological and Genetical Studies of the Interspecific Cross *Setaria Italica S. Viridis*," *K'o-hsüeh nung-yeh* 1, no. 4 (June 1944): 229–48; Li Ching-hsiung and Li Hsien-wen, "Cytological Studies of a Haploid Wheat Plant," *K'o-hsüeh nung-yeh* 1, no. 3 (March 1944): 183–89; and Li Hsien-wen and Chang Lien-kuei, "Maize Breeding Theory and Szechwan Hybrid Maize Breeding," *Nung-pao* [Farming] 13, no. 1 (February 1947): 3–13.

[42] Harlan, "Plant Breeding and Genetics," 301–4.

in the research literature of the Nanking geneticists were global in scope. In addition to recent American research literature, Russian studies of wheat and Japanese studies of rice are very prominent.

When Joseph Needham visited the Szechwan institute in 1943, he commented on its fine work and recorded a good-natured observation made by Li Hsien-wen: "A certain kind of barley, selected at the University of Nanking in 1937, was taken to the United States by Dr. Love, and proved to be so good that 'Wang Barley', [as it is called], has by now more than paid for all the expenses incurred by the United States in sending technical experts to China."[43] This anecdote summarizes well the successful development of genetics in China out of the American scientific tradition. But after 1949, the value of classical genetics was unacknowledged in China. The practical successes of Chinese geneticists were soon contested by the Chinese Communist Party and then denied as a matter of government policy. Eventually, the story of China's first modern geneticists was buried and lost under the subsequent decades of political oppression.

Epilogue: On Russians Bearing Gifts

In 1950, less than two years after it became Stalinist orthodoxy in Russia, Lysenkoism became a highly publicized and thoroughly enforced dogma in China. Reviling "bourgeois-idealist" Mendel-Morganism, the Lysenkoists denied the reality of the gene and most of the processes of heredity and variation formulated by Western and Russian geneticists over the previous forty years. Lysenkoists believed in the spontaneous generation of life, that inheritance factors are diffused throughout an organism's protoplasm, and that acquired characters are inheritable. They denounced mathematical and statistical methods of research, advocating the intuitive, trial-and-error methodology exemplified in the work of Michurin, a Russian horticulturalist.

In China, between 1949 and 1957, neither the geneticists nor any other experimental biologists trained in the West published in a Chinese journal, and few, if any, did research—unless it was in the Lysenkoist mode. Russian biologists lectured endlessly in Chinese schools, and their lectures became prescribed textbooks. Chinese biologists made pil-

---

[43] Needham and Needham, *Science Outpost*, 115.

grimages to Russia. Some were publicly reprimanded by Russians for their scientific errors. Those Chinese biologists who supported all this did so for a variety of reasons: Lysenkoist claims to rapid crop improvement; the simplicity of Lysenkoist science and its accessibility to the common farmer; and its philosophic claims of dialectical-materialist orthodoxy.

During the full bloom of the Hundred Flowers policy (fall of 1956), Li Ju-ch'i, T'an Chia-chen, and a score of their colleagues confronted and prevailed over the Lysenkoists at a conference in Tsingtao. Li, T'an, and others thereafter led the way back to the international scientific community, until they were again cut off in 1966.

Since 1977, genetics has been rehabilitated as a legitimate and even central area of scientific research. T'an Chia-chen and his colleagues enjoy a preeminent position in the Chinese scientific community. They are leading the way, under government sponsorship, in China's effort to achieve the most advanced international levels of contemporary genetics.[44]

---

[44] For genetics in China after 1949 see my *Lysenkoism in China: Proceedings of the 1956 Qingdao Genetics Symposium* (Armonk, New York, 1986); and my "Learning From Russia: Lysenkoism and the Fate of Genetics in China, 1950–1986," in *China's New Technological Revolution*, eds., M. Goldman and D. Simon (Cambridge, Mass., 1988).

Botany in Republican China:
The Leading Role of Taxonomy

William J. Haas

It is a moot question whether one group of men will succeed in recording the remaining unclassified organisms before other men succeed in inadvertently destroying them.[1]

A novelty of the 1930 Fifth International Botanical Congress in Cambridge, England, was a symposium on the flora of China. There were by then enough taxonomists versed in the flora of China to schedule a symposium. A botanist from China participated. In his address to the symposium, Ch'en Huan-yung (Woon-young Chun 陳煥鏞), the Harvard-trained leader of botany at National Sun Yat-sen University in Canton, divided the history of Chinese botany into three phases, "the period of ancient Chinese research, the period of early European research, and the period of modern Chinese research."[2] He focused his attention on

---

[1] Lincoln Constance, "Plant Taxonomy in an Age of Experiment," in *Fifty Years of Botany*, ed. William Campbell Steere (New York, 1958), 581.

[2] Chun Woon-young, "Recent Developments in Systematic Botany in China," in *Fifth International Botanical Congress, Report of Proceedings* (Cambridge, 1931), 524. For a brief description of the Fifth International Botanical Congress, see A. B. Rendle, "A Short

the last period. This had begun in 1916, when "the first Chinese botanist to describe new species of plants" published his descriptions.[3] When Ch'en listed the Chinese institutions emphasizing taxonomy from this time forward, he did not have to tell his colleagues that the list included all the major botanical institutions in China. The other participants in the symposium had first-hand knowledge; they themselves had contributed to the development of these institutions, thereby helping to make taxonomic research more prevalent than work in other fields of biology in China during the 1920s and 1930s. What institutional arrangements made this preeminence of taxonomy possible?

Modern botanical taxonomy was developed in China by Chinese adopting Western taxonomic practice and participating in a system of relations among Western institutions. Since taxonomic practice shaped institutional relations, it is described below as a basis for discussing these relations. Since the Chinese joined an existing system of institutional relations, relations which followed historically developed needs of collecting and distributing plant material, the pattern of collecting in the period before the Chinese joined in the work is also reviewed. After surveying China's centers for botanical work and the pattern of their funding, I will review the role of Harvard's Arnold Arboretum in the training of Chinese botanists, in the development of exchange relations, and in the financing of botanical exploration in China. Chinese institutions had relations with a number of their European and American counterparts. The arboretum achieved its outstanding role in Chinese botany due both to the interest in Chinese species of its first director, Charles Sprague Sargent, and, later, to the appointment of a leading expert on Chinese plants, Elmer Drew Merrill, as supervisor of all of Harvard's botanical institutions. The arboretum did not typify Western institutions working in Chinese botany, but its activities are representative of the interactions that took place between Chinese and Western institutions. Finally, the essay closes with a selective account of the disruption of research by the Sino-Japanese War. We watch the

---

History of the International Botanical Congresses," *Chronica Botanica* 1 (1935): 39–40.

[3] These descriptions by Ch'ien Sung-shu were published in the New England Botanical Club's journal, *Rhodora*. See Ch'ien Sung-shu, "Two Asiatic Allies of *Ranunculus pensylvanicus*," *Rhodora* 18 (1916): 189–90.

difficulties of Ch'en Huan-yung, Hu Hsien-su 胡先驌, and their institutions. Although the Republican period extends to 1949, the end of the war is the termination point of this study. Postwar research and operations did not recover prewar vigor. The years after the war and before Liberation are not the end of the present story, but the beginning of its sequel.

A description of cross-cultural contact requires detailed accounts of individuals' experiences. These experiences are the cross-cultural contact, the meat of the story. Nonetheless, my selection has been stringent. Most space goes to Ch'en Huan-yung and Hu Hsien-su, the outstanding leaders of the period. Ch'en opens and ends the story. He also makes several appearances in between, as does Hu Hsien-su. Others, such as Western expert Elmer Drew Merrill, foundation officer N. Gist Gee, and Harvard professor of dendrology John Jack, appear more briefly.

Taxonomic Practice and Institutional Communication

Systematists identify, name, and classify collections of specimens that represent the diversity occuring in nature. Systematic work—whether the identification of individual specimens, the preparation of monographs on a group of organisms, or studies which consider the totality of plant or animal life over a geographic area—requires access to the specimens and literature of all previous work. The working literature of taxonomy and the specimens referred to therein are dependent parts of one system of knowledge.

One group of specimens known as "types" (defined loosely as the specimens used by authors to describe previously unknown species) has a special status because of its use as the ultimate standard for applying names to unstudied material. Some specimens can be identified instantaneously by mere inspection. More questionable specimens must be studied by comparison with specimens whose identity is already known, including type specimens. Thus, literature providing the location of types is necessary for research to proceed.[4] Botanical taxonomists labor to

---

[4] For a general discussion of the usefulness of types in taxonomic work and discussion of specific cases concerning Chinese specimens, see Elmer Drew Merrill, "On the Desirability of an Actual Examination of Extant Types of Chinese Specimens," *Sinensia* 3 (1932): 53–62.

relate literature and specimens directly by producing reference works, adjusting nomenclature, and writing monographs. The economic and social organization necessary for preparing, storing, and transporting specimens and the preparation and publication of literature describing them are so great as to make taxonomic work, at any but the most primitive level, a social enterprise. Through affiliation with herbaria, botanic gardens, and natural history museums the individual botanist gains access to a worldwide network. Researchers can thus bring together materials they need without having to expend the resources necessary to collect and prepare them themselves.

All institutions exchange literature, but only those that study easily transportable material regularly exchange specimens. The system of exchange for botanical material is highly developed. It has features one finds in monetary systems—for example, the keeping of accounts and the use of standardized means for communicating information and material. Fairly standard methods for the preparation, preservation, description, and storage of plant specimens have all helped this system develop. These methods depend on the nature of the material.

While the preservation of zoological material often requires elaborate preparation, plants are easily preserved. In the best modern practice they are pressed, dried, and mounted with glue on standard-sized sheets of 100 percent rag content mounting paper. Herbarium sheets are filed in steel cases according to some standard system of classification, forming a library of specimens. Each sheet has a label, usually on the lower right-hand corner, giving the specimen's Latin name, the place of collection, the date, the collector's name, a number, and often further details about the appearance of the plant and its ecology.[5] Collectors assign a separate serial number to each group of different specimens, known as a collection, often making as many as twenty specimens for each number. Each number is taken from a single individual, if a tree, or a group of individuals of a population apparently of the same species, if an annual. Thus, collections can be divided up into multiple sets, each set having identical numbering.

---

[5] For descriptions of herbarium practice, see George H. M. Lawrence, *Taxonomy of Vascular Plants* (New York, 1951), chap. 11, 234–62; and Albert E. Radford et al., *Vascular Plant Systematics* (New York, 1974), chap. 31, 751–74. Both sources provide further bibliography.

Institutions exchange specimens worldwide in a way similar to the interlibrary loan system. This exchange of specimens is not limited to temporary loans but also includes permanent transfer of materials from one institution to another. An institution that sponsors collecting will exchange duplicate sets of its specimens for those of other institutions and keep accounts of credits or debits against these institutions. Though the precision of these accounts is astounding, there is often room for negotiating their settlement. For example, specimens can be offered in exchange for journals, books, and reprints, or the value of specimens can be adjusted in order to make accounts balance.

The Development of Exchange Relations and Collections

Taxonomic research is dependent on an international system of botanical exchange. Although we do not yet have a complete history of this system of exchange, we do know that as early as the sixteenth century individuals were exchanging dried specimens.[6] Large-scale exchanges and loans between institutions were certainly going on by the middle of the nineteenth century, a trend no doubt related to the increased reliability of transportation. Cooperation was also aided by use of standard practices for the handling and care of large research collections. For example, Elmer Drew Merrill, administrator of Harvard's botanical collections from 1935 to 1948, was concerned that herbaria cooperate by having strict programs for insect control, a problem central to the management of all herbaria. Because of his experience combating an infestation of herbarium beetles at the Arnold Arboretum, Merrill understood the constant danger of potential reinfestation from borrowed specimens, returned loan material, and duplicates exchanged with other institutions.[7] Merrill's own work depended on material sent from or through other institutions in remote locations, and his promotion of improved techniques and cooperation facilitated that work.

---

[6] A. G. Morton, *History of Botanical Science* (New York, 1981), 153, n. 26.

[7] Elmer Drew Merrill, "International Cooperation among Botanists," *Chronica Botanica* 1 (1935): 6; and "On the Control of Destructive Insects in the Herbarium," *Journal of the Arnold Arboretum* 29 (1948): 106.

Taxonomic work was often difficult in institutions in remote locations because of the need for material, especially type specimens. The location of types housed in the world's herbaria reflects the history of plant collecting. The great European herbaria accumulated North American material from the early seventeenth century, before botanical institutions were established in the New World. The presence of many New World types deposited in Old World herbaria was an impediment to the development of American botany. A new pattern of specimen acquisition began in the 1830s and 1840s. As part of their cooperative work on the flora of North America, John Torrey and Asa Gray, botanists at Columbia and Harvard respectively, acquired specimens from collectors within the United States for their own herbaria. American material thereafter was increasingly accumulated by American institutions.[8]

From the beginning of the twentieth century, systematic photography of American specimens in European herbaria also made work more convenient for American botanists. Harvard botanist M. L. Fernald was one of the first to use photography on a wide scale. From the early 1900s he photographed specimens in British herbaria, since his work required continued access to large numbers of specimens. In 1945, when working on the eighth edition of *Gray's Manual of Botany* on the flora of the northeastern United States, Fernald packed off his junior colleague, Bernice Schubert, to the British Museum (Natural History) and the Royal Botanic Gardens at Kew to photograph the needed specimens.[9]

The early collections of East Asian plants, which began to be amassed early in the eighteenth century, were also deposited in European herbaria. American researchers working on the flora of East Asia from the second half of the nineteenth century required access to the

---

[8] Radford et al., *Vascular Plant Systematics*, 28. For more on Gray's work with Torrey, see A. Hunter Dupree, *Asa Gray* (Cambridge, Mass., 1959); for the accumulation of types in Gray's herbarium, see Benjamin Lincoln Robinson, "Botany," in *The Development of Harvard University Since the Inauguration of President Eliot, 1869–1929*, ed. Samuel Eliot Morison (Cambridge, Mass., 1930), 349.

[9] M. L. Fernald and Bernice G. Schubert, "Studies of American Types in British Herbaria," *Contributions from the Gray Herbarium of Harvard University* 167 (1948): 149. This article was reprinted from *Rhodora* 50 (July 1948): 149–76; 50 (Aug. 1948): 181–208; 50 (Sept. 1948): 217–33.

material in European collections. As American institutions built up their own collections of East Asian materials, they began to participate in an exchange which became essential for taxonomic work on both sides of the Atlantic.

From the beginning of the eighteenth century to a high point in the early twentieth century, Western interest in the flora of China increased. The collection of live plants from China was fueled by Western commercial and academic entrepreneurs eager to exploit East Asian species for their horticultural and economic value. Private organizations, companies such as the British horticultural firm of Veitch & Sons, or joint enterprises organized for single ventures underwrote the expenses of collection in China in anticipation of substantial returns, either money or glory. This activity benefited taxonomic research, since dried specimens of each of the live plants were usually prepared. By forwarding these dried specimens to professional taxonomists, horticulturists obtained exact identifications of the plants they sold. While science served commerce, commercial activity served scientific ends by contributing to the accumulation of Chinese specimens in the research collections of Western herbaria.[10]

Collecting expeditions increased in the 1840s and 1860s after access to previously unexplored areas became possible.[11] The Treaties of Nanking in 1842 and Tientsin in 1858, after the first and second Opium Wars, compelled China to give foreigners more freedom to travel. By the early twentieth century, horticultural introductions from China were the rage in Europe and the United States. In the first three decades— the most fruitful period for bringing horticultural specimens back from China—roughly one third of trees and shrubs coming to the West from around the world were of Chinese origin.[12] While privately financed expeditions continued to collect in China, governments also began to

---

[10] For a general account, see E. H. M. Cox, *Plant-Hunting in China: A History of Botanical Exploration in China and the Tibetan Marches* (London, 1945).

[11] These increases can be seen by rearranging the summary of botanical exploration in China found in Emil Bretschneider, *History of European Botanical Discoveries in China* (London, 1898), 1077-90, from grouping by locality to grouping by date.

[12] W. W. Smith, "The Contribution of China to European Gardens," *Notes from the Royal Botanic Garden, Edinburgh* 16 (1931): 215-21.

participate. For example, the U.S. Department of Agriculture's Bureau of Plant Industry was sending its own collectors to China from 1905 on.[13]

There is a close relationship between the taxonomic literature and the specimens available for study. The large number of Chinese specimens sent to the West as a result of accelerated collecting from the mid-nineteenth century on was probably the main cause of increased publication. The number of Western authors writing on Chinese plants and the number of their publications grew especially fast towards the end of the nineteenth century and had a phenomenal rise in the 1920s and 1930s. The sharp increase in publication beginning in the 1920s reflected not only horticultural concerns but also botanical interest in the new diversity uncovered by everwidening exploration in China.[14] In the 1920s the whole enterprise was further stimulated when Chinese botanists began sending large collections of dried specimens to Western institutions.

Chinese Institutions Join the System

Beginning in the early 1910s, a few Chinese students had already come to Europe and America to study taxonomic botany. This was the only way they could take advantage of the large quantity of Chinese specimens collected by missionaries, diplomats, and traders and sent to Western institutions over the previous two centuries. Without access to the research collections, both specimens and literature, in Western herbaria, Chinese botanists would have had to begin work on the flora of their country from scratch. By 1916, when Western-trained Chinese

---

[13] For a general history of plant introduction by the U.S. Department of Agriculture, see the autobiography of David Fairchild, *The World Was My Garden* (New York, 1938). Isabel Cunningham provides a biography of the Department of Agriculture's main explorer in China from 1905 to 1918 in *Frank N. Meyer, Plant Hunter in Asia* (Ames, Iowa, 1984). For a general bibliography on plant introduction to the United States, see Margaret Rossiter, *A List of References for the History of Agricultural Science in America* (Davis, Calif., 1980), 17–18.

[14] My observations on publication are based on my study of Elmer Drew Merrill and Egbert H. Walker, *A Bibliography of Eastern Asiatic Botany* (Jamaica Plain, Mass., 1938).

botanists began to return to China to take up taxonomic work, the Western exchange system was already highly developed. Chinese botanists later participated in this system, thus taking advantage of continuing Western work on China.

In 1930, after a decade of Chinese activity, Western attention to Chinese plants as a special field within botanical taxonomy led to the symposium on the flora of China at the Fifth International Botanical Congress. It was this conference that included Ch'en Huan-yung's address on "Recent Developments in Systematic Botany in China," in which Ch'en took the opportunity to put on record his own name and the names and positions of China's four other leading botanical taxonomists. Of the five mentioned, four had been trained at Harvard University's Arnold Arboretum. Two of the four, Ch'ien Sung-shu (S. S. Chien) 錢崇澍 and Chung Hsin-hsüan (H. H. Chung) 鍾心煊, were recipients of Tsinghua scholarships in 1910 and 1911.[15] In 1908 the United States returned part of the indemnity China was forced to pay as settlement for claims from the Boxer War in 1900. These funds provided scholarships to Chinese for study in the United States. Additionally, Tsinghua College was established to prepare students for training abroad.[16] The remaining two trained at the Arnold Arboretum, Hu Hsien-su (H. H. Hu) and Ch'en, probably paid for most of their education in the United States with their own funds. They financed their study partly with Harvard loans. The fifth Chinese botanist, Ch'in Jen-chang (R. C. Ching) 秦仁昌, was a student of Ch'en and Ch'ien. After working as Hu's assistant and doing considerable collecting,[17] Ch'in received China Foundation funding from 1930 to 1932 to study ferns with the Danish botanist, Carl Christensen, and to photograph Chinese specimens of flowering plants and ferns in European herbaria.

---

[15] Tsinghua Alumni Association, *Alumni Yearbook, 1923* (Peking, 1923), 92, 95. The names of Ch'ien and Chung are given variant spellings, Chien Tsun-hsü and Chun Shin-hsün, respectively.

[16] For a brief history of the Tsinghua program, see *Tsinghuapper, 1924–25* (Peking, n.d.), 172–75.

[17] Ch'in dedicated volumes 3 and 4 of his *Icones Filicum Sinicarum* (Peking, 1935 and 1937) to his teachers Ch'ien Sung-shu and Ch'en Huan-yung. Hu Hsien-su mentions Ch'in as his assistant in 1924 in "Recent Progress in Botanical Exploration in China," *Journal of the Royal Horticultural Society* 63 (1938): 382.

Ch'in photographed Chinese types—as Fernald had photographed American types in European herbaria—and thereby made botanical work in China much easier. Further, duplicate sets of prints made from Ch'in's negatives were used to advantage by the Fan Memorial Institute of Biology in Peking as material for exchange with institutions, mostly American, whose access to European herbaria was also inconvenient. Ch'in's collection of photographs was unusually large, approximately 18,000 specimens.[18] The only larger contemporaneous collection was photographs of South American specimens prepared by Chicago's Field Museum of Natural History with funding from the Rockefeller Foundation, perhaps the foundation's only aid for a large taxonomic project.[19]

How did the work of Ch'en and his colleagues—Ch'ien, Chung, Hu, and Ch'in—mesh with international study of the flora of China? The training of Ch'en and his colleagues in the West and their return to work in China were significant features in the development of a part of Western botany; yet, from an equally cogent perspective, their activity in developing botany in China was a distinctive part of twentieth-century Chinese history. While China was for Western botanists only one of a number of geographic areas studied, it was the core of an entire discipline for the Chinese. Botany in China was almost exclusively the taxonomic study of Chinese flora.

Western researchers were willing, indeed, eager to cooperate with Chinese efforts, since research on the flora of China had acquired prominence in Western taxonomic botany during the early part of this century. Chinese taxonomists were returning home to an area at the forefront of Western exploration. They obtained financial support for their research in China by integrating their normal activities with the interests of Western institutions. Taxonomists working throughout China

---

[18] For the location of the collections photographed and the distribution of photographs, see Fan Memorial Institute of Biology, *Annual Report of the Fan Memorial Institute of Biology* 3 (1931): 3; 4 (1932): 5–6; 5 (1933): 16; and 6 (1934): 17. For Ch'in's funding, see the tables for science research fellowships in the China Foundation for the Promotion of Education and Culture, *Report* (Peking, 1930 and 1931). Brief descriptions of Ch'in's work can be found in the 1931 *Report* (p. 39) and the 1932 *Report* (p. 44).

[19] Hu Hsien-su, "Recent Progress of the Botanical Sciences in China," *Peking Natural History Bulletin* 9 (1934): 74; Rockefeller Foundation, *Annual Report* (New York, 1929), 232.

could interest foreign institutions in their work because they could collect and describe specimens using unique material unavailable in the West. Western botanists would have to take into account descriptions of species by workers in China because of the rule of priority. According to this rule, the earliest name used for a species, if in accordance with international rules, must always be used for that species.

Chinese biologists trained in the West in experimental biology lacked the level of financial support obtained by their taxonomist colleagues; most returned to China only to languish in positions with heavy teaching responsibilities. Often they would shift their research to an area of taxonomy. There were exceptions. Some geneticists flourished at Yenching University and the University of Nanking; these institutions had access to special funding from the Rockefeller Foundation and from the American Committee for China Famine Relief for genetics research to improve crop yields.[20] Other exceptions were physiologists who did research at the Rockefeller Foundation-supported Peking Union Medical College or at the Lester Institute in Shanghai.

Research opportunities for Western biologists in China, overwhelmingly American, followed the same pattern. A striking example of an experimentalist switching to a taxonomic field is Alice Boring. In 1923, Boring, sister of experimental psychologist Edwin Boring, took a position teaching biology at Yenching University in Peking. Although she had been trained at Bryn Mawr under the cytogeneticist T. H. Morgan and had coauthored with him her first published paper, she found little support for her experimental research when she reached China. She soon came to realize the value of doing taxonomic work in a previously unexplored area. After only a few years in China she shifted from cytogenetics to taxonomy of reptiles and amphibians. She began this work by preparing a checklist of Chinese amphibians and notes on their geographical distribution as junior coauthor with N. Gist Gee, a Rockefeller Foundation advisor in China.[21]

As Chinese botanists trained in the West began to return to China in 1916, a new type of Western botanist took up residence in China. Until the early 1910s most Western botanists in China were amateurs

---

[20] See Laurence Schneider's paper in this volume for particulars.

[21] N. Gist Gee and Alice M. Boring, "A Checklist of Chinese Amphibia with Notes on Geographical Distribution," *Peking Society of Natural History Bulletin* 4 (1929): 15.

(even if working at a professional level). There were also some professional collectors. In the late 1910s American botanists began to take up long-term residence in China as teachers in mission colleges and universities. Floyd Alonzo McClure, after receiving a master's degree from Ohio State in 1919, began teaching at Lingnan University in Canton. Later known as the "bamboo man" for his extensive studies on bamboos, McClure also worked for the U.S. Department of Agriculture as a plant explorer from 1923 to 1927. Another botanist, Albert Newton Steward, left Oregon State College with his bachelor's degree in 1921 and took a teaching position at the University of Nanking; and Franklin Post Metcalf began teaching at Fukien Christian University in 1923 after a stint with the U.S. Department of Agriculture. Each of the three stayed in China for approximately two decades, while other Western botanists, such as Egbert H. Walker, coauthor with Elmer Drew Merrill of the famous *A Bibliography of Eastern Asiatic Botany*,[22] stayed for shorter periods. The work of all these men helped shape the course of Chinese botany.

Newly arrived Western botanists, such as McClure, Metcalf, and Steward, and newly returned Chinese botanists—Ch'ien Sung-shu, Ch'en Huan-yung, Hu Hsien-su, and Chung Hsin-hsüan—had much in common, both in the range of their activities in collection, publication, and the setting up of herbaria and in the areas from which they could elicit financial support for their work. They needed support for personnel and new facilities to collect, prepare, and transport specimens, and to publish the results of research.

Research Centers in China

Because of their concentrations of institutions with large herbaria, Canton and Nanking in the early 1920s and Peking in the late 1920s became the three main centers for botanical research.[23] In Canton,

---

[22] Walker later authored *A Bibliography of Eastern Asiatic Botany*, supp. 1 (Washington, D.C., 1960).

[23] For general descriptions of herbaria in China, see F. P. Metcalf, "Herbaria in China," *Lingnan Science Journal* 9 (1930): 351–56, and 10 (1931): 101–11; and Li Hui-lin, "Botanical Exploration in China during the Last Twenty-five Years," *Proceedings of the Linnean*

Lingnan University (formerly, Canton Christian College) and National Sun Yat-sen University were both major institutions. Lingnan opened its herbarium with the help of Elmer Drew Merrill, already a leading authority on the flora of China, in 1916. In this paper I can scarcely detail the importance of Merrill, in his successive positions at the Philippines Bureau of Science, the School of Agriculture at the University of California, Berkeley, the New York Botanical Garden, and Harvard's Arnold Arboretum, with respect to the intellectual and institutional development of botany in China. His influence derived not only from his promotion of institutional ties but also from his extensive

---

Society of London 156 (1944): 35–37. Brief remarks on institutions doing botanical research are given in Hu Hsien-su, "Chung-kuo chin-nien chih-wu-hsüeh chin-pu kai-k'uang" 中國近年植物學進步之概況 [Progress of Chinese botany in recent years], *Chung-kuo chih-wu-hsüeh tsa-chih* 中國植物學雜志 [The Journal of the Botanical Society of China] 1, no. 1 (1934): 3–10. Information on individual herbaria and their collections is found in F. A. McClure, "A Brief Historical Survey of the Lingnan University Herbarium," *Lingnan Science Journal* 7 (1929): 267–90, for Lingnan University; Chou Lou, *The National Sun Yat-sen University: A Short History* (1937), 94, for National Sun Yat-sen University; S. T. Dunn, "The Hongkong Herbarium," *Kew Bulletin of Miscellaneous Information* 6 (1910): 188–92, for the herbarium of the Hong Kong Botanical Garden; College of Agriculture, *A Brief Statement of the College of Agriculture, National Southeastern University,* Information Bulletin No. 1 (Nanking, 1921), for National Central University; Science Society of China, *The Science Society of China: Its History, Organization and Activities* (Shanghai, 1931), 25–28, for the botany department of the Biological Laboratory; Nanking University, *Report of the College of Agriculture and Forestry and Experiment Station* (Shanghai, 1919–31) [before 1922 the title was *Report of the College of Agriculture and Forestry*]; Academia Sinica, *The Academia Sinica and Its National Research Institutes* (Shanghai, 1931), 117–25, for the Metropolitan Museum of Natural History (precursor to Academia Sinica's Institute of Botany and Institute of Zoology); Fan Memorial Institue of Biology, *Annual Report of the Fan Memorial Institute of Biology,* 1st through 6th issues (Peking, 1929–35) (there are additional reports I have not seen), and Hu Hsien-su, *The Fan Memorial Institute of Biology: An Outline of Its Work and Plans* (1935), for the Fan Memorial Institute. There are reports for the National Academy of Peiping that I have not seen.

identifications of specimens in Chinese herbaria. Expert at so-called "sight identifications" (instantaneous identification without reference to known material), Merrill was able to identify approximately 75,000 Chinese specimens from 1914 to 1929.[24]

Originally operated by an instructor in animal husbandry, the Lingnan herbarium began to grow more rapidly when Floyd Alonzo McClure arrived in 1919 and started vigorous collecting on Hainan Island (a large island situated south of Kwangsi Province). Another strong push in the development of the collections came when Franklin Post Metcalf arrived in 1932. From 1923 to 1929 Metcalf had taught at Fukien Christian University and set up a herbarium there. In 1928 the botanical laboratory of the other major university in Canton, National Sun Yat-sen University, was opened under the direction of Ch'en Huan-yung, who moved to Canton from his position at National Central University in Nanking. In 1934, in addition to his duties at Sun Yat-sen, Ch'en took on responsibility for the Research Institute of Biology at the University of Kwangsi. The accessibility of the excellent collections of the herbarium of the Hong Kong Botanical Garden (established in 1878) further enhanced the status of Canton as a botanical research center.

Northward in Nanking, the nation's capital beginning in 1927, four institutions had substantial collections: National Central University, the Biological Laboratory of the Science Society of China, the University of Nanking, and the Institute of Biology of Academia Sinica. Ch'ien Sungshu and Hu Hsien-su started collections at National Central when they returned to China in 1916 to take up positions at that university's predecessor, Nanking Higher Normal College. The founding of the Science Society's Biological Laboratory in 1922 increased the scope of activity of both Ch'ien and Hu. Both the botany department at the Biological Laboratory and the one at National Central retained Hu as head until he left in 1927 to become the head of botany at the new Fan Memorial Institute of Biology in Peking, at which time Ch'ien took over his positions in Nanking. The University of Nanking got its push when Albert Newton Steward arrived in 1921 and began to supervise

---

[24] Elmer Drew Merrill, "The Local Resident's Opportunity for Productive Work in the Biological Sciences," *Lingnam Science Journal* 7 (1929): 293. See R. Schultes, "Elmer Drew Merrill—An Appreciation," *Taxon* 6.4 (May, 1957): 89–101, for a brief overview of Merrill's career.

collecting. Shortly thereafter, specimens began to be accumulated and the makings of a herbarium developed. In 1922 Merrill did for the University of Nanking what he had done for Lingnan University by helping to organize a herbarium. In 1930 the Metropolitan Museum of Natural History of the Academia Sinica opened in Nanking. Reorganized in 1934 as the Institute of Biology, its botanists were mainly students of Hu, Ch'ien, and Ch'en. The small herbarium of Ginling College (Chin-ling nu-tzu wen-li hsüeh-yuan) also added to the collections in the Nanking area.

Further north in Peking were the Fan Memorial Institute of Biology, the Institute of Botany of the National Academy of Peking, National Peking University, and National Peiping Normal College. The Fan Memorial Institute, established in 1928, rapidly became a major center for organizing exploration. The Institute of Botany was established in 1929. The two universities were said to have research collections. In contemporary accounts Peking University was considered the first school in China to have had a Chinese botany professor. Hu Hsien-su, the preeminent botanist in Republican China, discounted the work of this man as being in the older herbalist tradition.[25]

Besides these three major centers, strong collections also existed in Fukien, a coastal province opposite the island province of Taiwan. In addition to the herbarium set up by Metcalf at Fukien Christian University, Chung Hsin-hsüan set up a herbarium in 1922 at the University of Amoy. Again, Merrill played a major role in identifying the specimens of the collections at Fukien Christian University and the University of Amoy as he had done at Lingnan University and the University of Nanking. By 1929 the herbarium of Fukien Christian University had accumulated fourteen thousand specimens, of which ten thousand had been identified by Merrill.[26]

---

[25] Hu Hsien-su, "Recent Progress in Botanical Exploration in China," 381.

[26] See F. P. Metcalf, "The Herbarium at Fukien Christian University," *Proceedings of the Natural History Society of Fukien Christian University* 2 (1929): 26; and H. H. Chung, "The Study of Botany in Fukien," *Lingnan Science Journal* 7 (1929): 126–30.

Funding Patterns

This quick survey of botanical institutions mentioned some of the key personnel responsible for their development and operation. Merrill, an expert residing outside of China, contributed to the organization of herbaria. McClure, Metcalf, and Steward were new professional botanists from the United States who took up long-term residence in China. Ch'ien Sung-shu, Hu Hsien-su, Ch'en Huan-yung, and Chung Hsin-hsüan were the early "returned students," all with graduate training at Harvard's Arnold Arboretum. They were responsible for training the leading botanists of the next generation.

Funding for the institutions with which these men were associated came mainly from the Rockefeller Foundation, the China Foundation for the Promotion of Education and Culture, Chinese philanthropists and philanthropic organizations, and foreign research institutions. The various sources of funding played different roles in the development of botany in China. Funds for research would not have enabled botanists' efforts if there were not also funds for facilities and personnel. Institutions' needs were diverse and so, too, were the motives of funders. A single source gave funds to achieve its own limited goals, but, when put together, the money from a variety of funding sources made research possible.

To insure enough qualified applicants for its Peking Union Medical College, the Rockefeller Foundation opened its own premedical school in China in 1917. As a further measure, funds were granted to mission schools for establishing or improving science education. The failure of these measures to provide more than a handful of students for its $8-million college, however, led the Rockefeller Foundation to establish a formal premedical education program in 1922. Appropriations as a part of this program were crucial in developing science teaching at more than half the Christian colleges and universities in China and at some of the key Chinese universities as well. It should be emphasized, however, that Rockefeller Foundation funds did not cover research expenses; funds for buildings, equipment, and salaries for teachers constituted the bulk of appropriations. Although science teachers in colleges and universities could obtain Rockefeller aid to pursue advanced training or research for a year in China or abroad, such aid was intended as a means to improve teaching, not to encourage research. Receiving this support was contingent upon the teacher's agreeing to return to his teaching position at the end of the year. Leading Chinese biologists wanting to change their situations or who already had succeeded in removing themselves from teaching were not eligible for this aid.[27] In short, Rockefeller funding was not for researchers who were not also teaching.

In a 1933 report on scientific institutions in China, W. E. Tisdale, a Rockefeller Foundation officer from the European office, bemoaned the fact that Rockefeller aid had encouraged an emphasis on the taxonomic phase of biological research. But Tisdale's assessment was wrong. While the Rockefeller Foundation had given money to biologists who engaged in taxonomy, it did so without regard to these biologists' research interests. Money was disbursed to teachers. A large number of foreign teachers in China were involved in taxonomic work, but their decision to do this kind of work was made without regard to Rockefeller aid. Tisdale was unhappy about the predominance of taxonomy because of his inability or unwillingness to consider taxonomy a scientific discipline with stature equal to that of the experimental sciences. His report was also inaccurate because he was guided by the assumption that the Rockefeller Foundation was philanthropically omnipotent—that taxonomy could not flourish in China unless the Rockefeller Foundation wanted it to.

Nonetheless, in his own way Tisdale realized the limits of Rockefeller power. He thought the experimental sciences were not developing in China—despite the presence of Chinese professors with foreign training—because of a lack of Chinese with ability to inspire and direct their countrymen. To fill this need China required foreign professors. But "the comparative lack of immediate scientific opportunities in China" made it difficult to find Western scientists willing to make the

---

[27] Here I am building on Laurence Schneider's thesis, presented in "The Rockefeller Foundation, the China Foundation, and the Development of Modern Science in China," *Social Science and Medicine* 16 (1982): 1217–21. My remarks are based on reading correspondence and memos on the Rockefeller Foundation premedical education program in the records of the China Medical Board and the China Medical Board, Inc., at the Rockefeller Archive Center. For a summary of the premedical education program, see Catherine Lewerth, "A Draft History of the Rockefeller Foundation" (unpublished ms., Rockefeller Archive Center [hereafter, RAC]), 2946–3005.

sacrifice of isolating themselves.[28] Tisdale's solution was to get "Jews evicted from Germany" to go to China. But he realized that here the Rockefeller Foundation could not take the initiative. Rather, "[Chinese] Ministries and University authorities should be the active ones in arranging contracts."[29]

On the other hand, N. Gist Gee, advisor for the Rockefeller premedical education program and a primary figure in the development of natural history work in China, was aware of Western taxonomists' interest in China's flora and fauna, of growing Chinese excellence in this work, and of the Rockefeller Foundation's opportunity to take the lead in encouraging this work in China. Although he tried to push the foundation to fund directly projects that would enhance natural history in China, he was unsuccessful. The Rockefeller Foundation had no policy for the development of scientific research of any kind in China outside the Peking Union Medical College. The development of taxonomy at institutions receiving Rockefeller funding was a fortuitous result of the foundation's premedical education program. Lingnan University, National Central University, the University of Nanking, and Fukien Christian University were able to take advantage of opportunities for botanical exploration and research provided by foreign institutions because sizable portions of the expense for physical plants and salaries were covered by funds from the Rockefeller Foundation. Although Rockefeller funding made it possible for these institutions to do taxonomic research, other funding sources were also required.

The Rockefeller Foundation had developed a bias favoring reductionist biology that culminated in 1933 with a program headed by Warren Weaver for funding biological research informed by reductionist philosophy. Weaver argued many times with Ernst Mayr, already a leading ornithologist and systematist, "over support for systematic work but Weaver was insistent that the Rockefeller Foundation [would] only support breakthrough fields."[30] Like many other scientists, Weaver held that a reductionist approach was the way to make great innovations in biology. Weaver later admitted

---

[28] W. E. Tisdale, "Report of Visit to Scientific Institutions in China" (unpublished ms., 1933; RAC: RG.1/60l/40/326), 5.

[29] Ibid., 15.

[30] Telephone conversation with Ernst Mayr, 6 June 1984.

an overenthusiasm for those aspects of biological research that yield especially well to quantitative and precise measurement and to strict analysis. This makes it hard to be very excited over, say, systematics, even though one realizes the necessity for the collection and ordering of all these items of fact about nature.[31]

This reductionist bias, including, as it does, a strong antipathy to taxonomy, prevented Rockefeller officials like Tisdale and Weaver from seeing the development of taxonomy in China, a field using a non-reductionist approach, as important to science internationally or domestically.

By contrast, the China Foundation for the Promotion of Education and Culture, the institution charged with distributing funds from the second U.S. Boxer Indemnity remission, provided funds for studying the flora and fauna of China. Still, much of the aid was earmarked for institutional overhead rather than research. Rockefeller and China Foundation funding for overhead complemented one another. First, the China Foundation did not begin disbursing funds until late in 1925, approximately three years after the Rockefeller Foundation formally established its premedical program. Niches already filled by Rockefeller money did not have the capacity to absorb China Foundation money— or at least not as much of it. Second, the China Foundation program concentrated on Chinese institutions; the Rockefeller Foundation program concentrated on missionary institutions. Third, because the China Foundation board of trustees included members tied to the Rockefeller Foundation—notably, Roger S. Greene, resident director of the Rockefeller China Medical Board—the programs of the China Foundation could be coordinated to some extent with those of the Rockefeller Foundation.

Although not a member of the China Foundation board, N. Gist Gee, the Rockefeller premedical education advisor, was involved in its operation. The board required information on potential recipients of funds, and it was natural for Roger Greene to tap his subordinate Gee, who was already gathering this information for the Rockefeller Foundation. Since the board included Chinese such as geologist Ting Wen-

---

[31] Warren Weaver, "A Quarter Century in the Natural Sciences," in *The Rockefeller Foundation: Annual Report* (New York, 1958), 37.

chiang (V. K. Ting), who would be sympathetic to proposals involving study of China's natural resources, Gee's suggestions were more in tune with this board in Peking than with Rockefeller Foundation officers in New York. Thus, Gee had substantial opportunity to encourage the development of natural history. He was involved in the negotiations leading to the creation of the Fan Memorial Institute of Biology, a major project jointly funded by the China Foundation and the "Aspiration Society" (Shang-chih Hsüeh-hui), a Chinese philanthropic society. Since the Rockefeller Foundation had decided not to support the Fan Memorial Institute, Gee was allowed to join the new institution's board of directors; his presence would not indicate special Rockefeller Foundation interest or commitment.[32]

In addition to funding such research institutes as the Fan Memorial in Peking and the Science Society's Biological Laboratory in Nanking, the China Foundation introduced a measure of stability to university science education. The largest program was the China Foundation Science Professorships, a program which ran from 1926 to 1936 and involved six universities: National Central, Peiping Normal, Northeastern, Wuhan, Chengtu, and Sun Yat-sen. In each university, chairs were established in physics, chemistry, botany, zoology, and educational psychology for a period of six years. The professor was granted a year's "furlough" after six years' service. This program was crucial in maintaining key personnel in a period when university professors' salaries were often in arrears. At least half the chairs for botanists went to taxonomists.[33]

Another China Foundation initiative that helped to maintain personnel was the more prestigious Science Research Professor Program. Ch'en Huan-yung, the Harvard-trained botanist at National Sun Yat-sen University, was the recipient of a research professorship after holding a regular science professorship from 1929 to 1934. Later, Hu Hsien-su,

---

[32] See "Interview with Chou I-ch'un (Y. T. Tsur)周貽春," 12 January 1928; "Interview with Hu Ching-fu (C. F. Wu)胡經甫," 2 March 1928; "Interview with Liu Ch'ung-le (C. L. Liu)劉崇樂," 3 March 1928; and N. Gist Gee, letter to R. M. Pearce, 11 May 1928 (RAC: RG.4/45/1044).

[33] See the annual *China Foundation Report* for year-by-year data on this program.

also Harvard-trained, received a research professorship while serving as head of the Fan Memorial Institute.

The China Foundation's funding of the Fan Memorial Institute should be considered part of a continuing commitment that began in 1926, when the foundation aided work on the flora and fauna of China at the Science Society's Biological Laboratory. This commitment is also reflected in appropriations given to other institutions for exploration and collecting. Unlike funding from the Rockefeller Foundation, China Foundation funding accounted for part of the growth of taxonomy in China, a growth that eclipsed that of other biological fields.

The Role of the Arnold Arboretum

Leading institutions in China that maintained herbaria established relationships with Western centers for botanical research, including the Royal Botanic Gardens in London and Edinburgh, the Botanical Museum of Berlin, the University of California at Berkeley, the U.S. National Herbarium at the Smithsonian in Washington, D.C., the New York Botanical Garden, and Harvard University's Gray Herbarium and Arnold Arboretum. Of all these institutions, the Arnold Arboretum had the greatest overall impact on the development of taxonomic botany in China. It attracted and trained the top men in the early years of botany in China, both Chinese and Western, and it was central in the exchange relations developing between Chinese and Western institutions. Why should Harvard have become such a center, and what was the scope of its exchange relations with China? Further, how was the arboretum's relationship with two Chinese institutions, the botany departments of National Central University and the Fan Memorial Institute of Biology, influenced by Hu Hsien-su?

The Arnold Arboretum, located on a 265-acre site in West Roxbury's Jamaica Plain, about five miles from the center of Boston, was established in 1872 as Harvard University's museum of living trees. In 1882, Harvard agreed to a restricted use of the arboretum as a part of the Boston park system in exchange for city funds for laying out the grounds. The arboretum was designed as part of the city park system by landscape architect Frederick Law Olmsted. Olmsted had been inspired to combine the functions of arboretum and park by Charles Sprague Sargent, director of the arboretum from 1873 until his death in 1927.[34] During this period Sargent established the arboretum as a center for the study of woody plants of the world.

The basis of Sargent's horticultural interest in East Asian species was Asa Gray's work on the similarity of Japanese and northeastern American plants. Gray accounted for this similarity by the theory that the floras of eastern Asia and eastern North America were derived from the same source. To the horticulturist, this significantly implied that the species of one location might grow well in the other. If Gray's work provided the theoretical underpinning for Sargent's horticultural interest, it was the flourishing of seeds sent to Sargent by Emil Bretschneider, a Russian physician in Peking, which gave Sargent the practical demonstration that collecting in China could provide viable plants of horticultural value.[35]

Although Sargent's collecting of Chinese species began slowly, with acquisitions through European institutions and a trip of his own to China, by 1907 he had hired E. H. Wilson from the English horticultural firm of Veitch & Sons. Wilson's fabulous collections from western China made him and the arboretum world-famous. Later, Sargent obtained the services of another famous collector and ethnologist, Joseph Rock.[36] Through the 1920s and 1930s the arboretum's herbarium increased its collections of plants from China more rapidly than those

---

[34] C. S. Sargent, "The First Fifty Years of the Arnold Arboretum," *Journal of the Arnold Arboretum* 3 (1922): 130; and Cynthia Zaitzevsky, *Frederick Law Olmstead and the Boston Park System* (Cambridge, Mass., 1982), 58–64.

[35] For a historical summary of the literature on the similarity between the flora of eastern Asia and eastern North America, see Li Hui-lin, "Floristic Relationships between Eastern Asia and Eastern North America" (Philadelphia, 1971), reprinted from *Transactions of the American Philosophical Society*, n.s. 42 (1952): 372–73; and D. E. Boufford and S. A. Spongberg, "Eastern Asia–Eastern North American Phytogeographical Relationships—A History from the Time of Linneaus to the Twentieth Century," *Annals of the Missouri Botanical Garden* 70 (1983): 423–39. For Gray's work in this area, see Dupree, *Asa Gray,* chap. 13. A summary of Bretschneider's shipment of seeds to Sargent is contained in a letter from Bretschneider to Sargent, 25 September 1893 (Arnold Arboretum Archive, Harvard University).

[36] See Stephanie Sutton, *Charles Sprague Sargent and the Arnold Arboretum* (Cambridge, Mass., 1970), chaps. 8–10.

from any other area.[37] But it was not just acquisition of Chinese collections that made the arboretum an important center. The investigation of these collections, especially by expert Alfred Rehder, an assistant at and later curator of the arboretum's herbarium, also contributed to knowledge of the flora of China.

Chinese students no doubt wanted to go to the arboretum because they were aware of its importance in research on the flora of China. Ch'en Huan-yung probably found out about the Arnold Arboretum through his teachers at the Massachusetts Agricultural College and through Marion Case of Weston, Massachusetts. Ch'en arrived from Shanghai in 1909 and enrolled in courses in forestry and entomology at the agricultural school in Amherst. In 1910 Ch'en was hired by Case as a summer assistant. For five summers between 1910 and 1919, he helped her train the boys employed at Hillcrest, Case's experimental farm in Weston, now the Case Estates of the Arnold Arboretum. The boys liked Ch'en, according to Case's stereotyped impressions, because the "quiet courteous ways he had inherited from his Spanish mother appealed to them."[38] Through a series of lectures given by botanists connected with the Arnold Arboretum (John Jack, an assistant professor of dendrology working under Charles Sargent, was a lecturer there), Hillcrest became a local center for diffusing horticultural knowledge.[39] These occasions may have provided an opportunity for Ch'en and Jack to become acquainted. Whether they originally met at Hillcrest during Ch'en's first summers in the United States or later, after Ch'en arrived at the Arnold Arboretum, their friendship was certainly stimulated by a mutual interest in the flora of China; Jack had gone to China in 1905 at his own expense to collect for the Arnold Arboretum.[40]

In 1912 Ch'en transferred from the Massachusetts Agricultural College to the New York State School of Forestry at Syracuse University. After graduating in 1915, he enrolled at Harvard's Bussey

---

[37] This can be seen by following the annual reports on the Arboretum's herbarium in the *Journal of the Arnold Arboretum*.

[38] Marion Roby Case, *The Second Summer at Hillcrest Farm* (Weston, Mass., 1911), 6.

[39] Sheila Geary, "The History of the Case Estates" (unpublished ms., in the author's possession), 3–4.

[40] John G. Jack, "The Arnold Arboretum: Some Personal Notes," *Chronica Botanica* 12, nos. 4–6 (1948–49): 187.

Institution for Research in Applied Biology. The Arnold Arboretum did not officially offer instruction, but students could arrange to take courses with John Jack and work at the arboretum by registering at the Bussey. Jack was good at teaching and his students adored him. He went out of his way to help them, often paying their wages for work at the arboretum out of his own pocket or arranging Harvard loans for them. This he did for Ch'en in 1916.

While there were more than a dozen Chinese pursuing graduate studies in various Harvard departments, Ch'en seemed more adventurous than most. Unlike his compatriots who resided in graduate dorms or right near school, Ch'en lived first on St. Botolph Street and later on Gainsborough Street in an "artsy" section of Boston's Back Bay—about a stone's throw from the Massachusetts Horticultural Society, Symphony Hall, and the New England Conservatory. Ch'en did well in his studies and in his final year, 1919, received one of Harvard's Sheldon Traveling Fellowships to collect plants in southern China. Harvard insisted on tapping Ch'en's Sheldon Fellowship for immediate repayment of his Harvard loan, but the dean of the Bussey Institution, entomologist William Morton Wheeler, interceded on Ch'en's behalf.[41] Wheeler was conducting his own worldwide taxonomic study of ants (this later included ants of China sent to him by Rockefeller advisor N. Gist Gee and others) and probably appreciated the value of having Ch'en collect Chinese plants for the arboretum.[42]

Since Sargent wished to expand his program for acquiring Chinese plant material, it made sense for him to utilize Chinese students trained at the arboretum as collectors after they returned to China. Ch'en Huan-yung was one of the earliest students to return to China in this role. I am not sure Sargent calculated this course of events, but it is exactly what came to pass.

---

[41] For Ch'en's addresses, see his Bussey Institution Registration and Record Card (UA V252.276, Harvard University Archives). For Wheeler's action, see Bursar, Harvard University, letter to John G. Jack, 23 May 1925 (Arnold Arboretum Chinese Correspondence, located at the Gray Herbarium, Harvard University [hereafter, AACC]).

[42] For Wheeler's publications on the ants of China, see the years 1921, 1923, 1927–29, 1931, and 1933 in the bibliography in Mary Alice Evans and Howard Ensign Evans, *William Morton Wheeler, Biologist* (Cambridge, Mass., 1970).

Unlike Ch'en, Hu Hsien-su returned to China for seven years between finishing his undergraduate degree at the University of California at Berkeley in 1916 and starting graduate training at Harvard in 1923. Hu's first direct contact with the Arnold Arboretum was through correspondence with Sargent. In 1920 he sent Sargent a collection of woody specimens from Kiangsi in exchange for their identification.[43] In this way Hu was able to build up his own research collection, attaching Sargent's identifications to an identically numbered duplicate set he retained. Sending specimens was also a standard way of making contact with eminent botanists. For example, as an undergraduate, Merrill put himself in touch with the director of the New York Botanical Garden, Nathaniel Lord Britton. In the same way as Hu did with Sargent, Merrill sent Britton a collection of Maine plants in exchange for literature.[44] Hu probably sent Sargent specimens more than once and no doubt began a working relationship with the arboretum years before he studied there.

Hu enrolled at the Bussey Institution from September 1923 to June 1925 and took four forestry courses with John Jack.[45] Like Ch'en, he borrowed money from Harvard, but he could not get as much as Ch'en had since, as Jack explained, Ch'en had not repaid his loan and this delinquency affected Hu's ability to secure funds. But Ch'en's tardy repayment may have been due as much to Harvard's negligence as to Ch'en's carelessness and difficult financial position. Jack was riled when a Harvard officer intimated that he (Jack) had "backed up a 'crook' for scholarships & other favors from the college."[46] This episode made Hu realize that Chinese students at the arboretum, and at Harvard University generally, were seen as a group. Hu heard about this problem from Jack only after he had returned to National Central University in Nanking. Ch'en, also teaching at National Central, immediately repaid half the loan, but he needed Hu's help to repay the rest since his salary

---

[43] Hu Hsien-su, letter to C. S. Sargent, 17 December 1920 (AACC). This correspondence is uncatalogued and unprocessed material.

[44] Elmer Drew Merrill, "Autobiographical: Early Years, the Philippines, California," *Asa Gray Bulletin*, n.s. 2 (1953): 344.

[45] For information on Hu's enrollment, see his Bussey Institution Registration and Record Card (UA V252.276, Harvard University Archives).

[46] John Jack, letter to Ch'en Huan-yung, 30 May 1925 (AACC).

was eight months in arrears. Hu raised the money to repay the balance of the loan "in anxiety of his [Ch'en's] error which may cast an ugly shadow upon the character of Chinese students at Harvard."[47] Hu was especially sensitive to the danger of Harvard attitudes towards Chinese that were based on negative stereotypes. His determination to clear Ch'en's debt perhaps grew out of his desire to promote a positive image of Chinese and their culture.

When Hu arrived at Harvard in 1923, he had already made a name for himself in China as one of a small goup of conservative intellectuals who were trying to preserve China's "national essence," that is, elements of the country's traditional culture. Hu's activity centered around publication of the magazine *Hsüeh-heng* [Critical review] with his colleagues at National Central University, Wu Mi and Mei Kuang-ti, who had been trained under the Harvard humanist Irving Babbitt. Babbitt wanted reformers "to retain the soul of truth that is contained in its [China's] great traditions" and no doubt reinforced the cultural conservatism of Hu and his colleagues.[48] Hu had had some contact with Babbitt when he was at Harvard and later translated an essay of Babbitt's views on humanities education for *Hsüeh-heng*.[49] This group of intellectuals expressed its conservatism by opposing literary reform, which called for use of the spoken language in place of the literary language as the written medium.[50]

---

[47] Hu Hsien-su, letter to John Jack, 2 October 1925 (AACC).

[48] H. C. Meng, "The New Literary Movement in China," *The Weekly Review of the Far East* (Shanghai) 20, no. 7 (15 April 1922): 250, quoted in Chow Tse-tung, *The May Fourth Movement: Intellectual Revolution in Modern China* (Cambridge, Mass., 1960), 282, note k.

[49] Wu Mi mentions Hu's delivery of a letter by him to Babbitt in a letter from Wu to Babbitt, 2 December 1925 (letters to Irving Babbitt, Babbitt Papers, HUG 1185.5, Harvard University Archives). For Hu Hsien-su's translation, see Pai-pi-te, "Pai-pi-te chung-hsi jen-wen chiao-yü t'an" [Babbitt's discussion of humanities education in China and the West], *Hsüeh-heng* [Critical Review], 1922, no. 3.

[50] On the "national essence" movement and the Critical Review group at National Central University—including the group's opposition to literary reform and the influence of Babbitt—see Laurence Schneider, "National Essence and the New Intelligentsia," in *The Limits of Change: Essays on Conservative Alternatives in Republican China*, ed. Charlotte Furth (Cambridge, Mass., 1976), 57–89; Bonnie

One reason some Chinese and Westerners promoted Chinese literary reform, part of a Republican-period "cultural revolution," was that they considered the literary language inadequate for scientific communication. From their point of view it likely seemed ironic that scientific research was promoted with great success at National Central University, a hotbed of conservative opposition to literary reform.[51] Of course, for Hu Hsien-su there was no contradiction between promoting Chinese botanical science—an activity he traced through the traditional period down to the Ch'ing-dynasty scholar Wu Ch'i-chün—and opposing literary reform; he was at the forefront of both activities.[52] Perhaps a lingering identification with conservative attitudes towards culture was, at least in part, what caused Hu to run afoul of China's post-1949 Cultural Revolution and be "cruelly treated by Lin Piao and the Gang of Four."[53] In the Republican period, Hu, along with many other Chinese scientists, published in *K'o-hsüeh* [Science] using the literary language. Even so, the debate over the use of the Chinese vernacular as opposed to the literary language as the written medium had no relevance to taxonomic work; the majority of Chinese publication in taxonomy was in English. Hu must have considered English appropriate for a reference work. He made no effort to make a Chinese translation of his Harvard doctoral

---

S. McDougall, *The Introduction of Western Literary Theories into Modern China, 1919–1925*, East Asian Cultural Studies Series, nos. 14–15 (Tokyo, 1971), 41–46; and Chow, *The May Fourth Movement*, 282.

[51] For the insights of Chao Yüan-jen 趙元任 on this situation, see *Yuen Ren Chao: Chinese Linguist, Phonologist, Composer and Author* (Berkeley, Calif., 1977), chap. 3, 53–54. This is a transcript of Rosemary Levenson's interview with Chao, made with the assistance of Laurence Schneider for the Regional Oral History Office of the Bancroft Library at the University of California–Berkeley.

[52] See Hu Hsien-su, "Fa-k'an-tsu" 發刊詞 [Foreword], *Chung-kuo chih-wu-hsüeh tsa-chih* 中國植物學雜誌 1.1 (1934): 1–2; and "Recent Progress of the Botanical Sciences in China," 71.

[53] Hu Hsien-su hsien-sheng chih-sang wei-yuan-hui [Hu Hsien-su's funeral committee], "Fu-kao" ["Obituary notice"], 15 May 1979. Located in the Gray Herbarium Library, Harvard University. No doubt Hu's wartime connection with the Kuomintang, as well as his opposition to Lysenkoist biology during the 1950s, also made him a target.

dissertation, a manual of the Chinese flora, a project he chose "in order to furnish Chinese students of botany a reference book to their native flora."[54]

As with the Chinese students, awareness of Harvard's importance as a place to work on the flora of China also accounts for study there by the three American botanists mentioned earlier, Metcalf and McClure for postgraduate work and Steward for his doctorate. The Arnold Arboretum played a large part in the work of all these men in China. In their institutional roles they called on the expertise and resources of the arboretum when identifications were needed, participated in exchange relations, and sought financial support for collecting expeditions.[55] Of the institutions mentioned earlier, all had substantial dealings with the arboretum except National Peking University, Peiping Normal University, and the Institute of Botany of the National Academy of Peiping. If the arboretum's financial support was relatively small when compared with Rockefeller and China Foundation appropriations, it should be noted that institutions in China did receive aid from other institutions in the West and, further, that this aid was often what kept exploration and research going. For example, in 1927, when Hu Hsien-su made more time for research work by removing himself from teaching at National Central University and taking his entire salary from the Science Society's Biological Laboratory, he maintained his exploration schedule by relying on arboretum funds from Sargent.[56] This support made the difference between being merely established in a position and being able to conduct vigorous research programs.

The full range of relations between Western institutions and institutions in China is beyond the scope of this article; indeed, the relations of just one key institution, the Arnold Arboretum, and the institutions in China it supported or exchanged with are too broad to relate fully. Table 1 shows figures for the total number of specimens exchanged

---

[54] Hu Hsien-su, "Synopsis of Chinese Genera of Phanerogams with Descriptions of Representative Species" (Ph.D. diss., Harvard University, 1925), 1.

[55] This can be documented by correspondence in the Arnold Arboretum Chinese Correspondence (AACC). In some cases, substantial institutional relations did not develop until the 1930s.

[56] Hu Hsien-su, letter to Alfred Rehder, 13 September 1927 (AACC).

Table 1

Specimens Exchanged between Institutions and the Arboretum before 1934

Institution	To Arnold Arboretum	From Arnold Arboretum
Fan Memorial Institute	8,535	1,976
National Central University	5,168	1,855
National Sun Yat-sen University	2,581	1,927
Metropolitan Museum	2,152	1,229
University of Amoy	1,381	1,743
Wuhan University	1,279	360
Lingnan University	1,270	734
University of Nanking	596	--

Source: Arnold Arboretum Exchange Records, no. 48 (Gray Herbarium Archives, Harvard University).

between all participating institutions in China and the arboretum up to the beginning of 1934.

The largest figures are for Hu's departments at National Central and, later, the Fan Memorial Institute. Of the 5,168 specimens sent to the arboretum from National Central, 4,868 were sent before Hu left for the Fan Memorial Institute in 1928. The figure for the Fan Memorial, 8,535, is all the more remarkable considering that it covers somewhat less than six years of operation. Hu Hsien-su was especially successful in obtaining arboretum support for the institutions he served, and these institutions had the highest level of exchange activity with the arboretum.

When he returned to Nanking in 1925 after spending the previous two years studying at the Arnold Arboretum, Hu immediately recruited John Jack to petition Sargent for arboretum support for collecting. After getting a pledge of funds from Sargent, Hu gratefully assured Jack that his confidence in the Chinese was not misplaced and that "besides our personal relation, the reputation of our nation and our race is in a way staked in this affair."[57]

Hu's deep gratitude should not obscure the fact that Sargent's assistance was not especially generous. Arboretum support came at a high price; the collecting institution was allowed to retain only two duplicate sets of specimens out of a total of ten or more sets collected. A contrast to Sargent's policy is the more liberal one instituted by Merrill at the New York Botanical Garden from 1929 and after he took over botany at Harvard in 1935: a fifty-fifty split of material between the collecting institution and the sponsor.[58] Merrill had a special sympathy for his colleagues working in China; he had had his own experience with Sargent's penury. In 1917 Sargent financed Merrill to collect in southern China. The understanding was that Merrill would make several sets of specimens for Arnold Arboretum distribution while retaining only

---

[57] Hu Hsien-su, letter to John Jack, 26 February 1926 (AACC). By the time Hu wrote this letter he had apparently obtained two grants from Sargent, each for six hundred dollars.

[58] For example, see Merrill's offer to Ch'en Huan-yung in a letter of 24 December 1930 (Merrill Correspondence, China; New York Botanical Garden Archives).

one for himself.⁵⁹ Merrill, like Hu, accepted an ungenerous offer from Sargent. Unlike Hu, in 1917 Merrill did not have to face competition from a number of institutions in China for limited Western support.

Hu was sensitive to the competition between institutions in China for Western support and between Western institutions for botanical material from China. Having obtained a positive response from Sargent for arboretum support of expeditions to Anhwei, Hunan, and Szechwan, he still worried about competition for these grants from the dean of the University of Nanking's School of Agriculture and Forestry, John H. Reisner. Writing anxiously to Jack, Hu revealed the extent of his concern.

> Reissner [sic] is outrageous in his act of competing with us for the financial support from the Arnold Arboretum, because he always professes friendship and cooperation with us, especially because he learned this news of Prof. Sargent's intention of financial support for botanical work in China from me . . . this attempt is nothing short of alienation of good intention between these two institutions, an act which must be condemned as immoral. If this is Christian spirit, no wonder our young men now endeavor to spread a national-wide anti-Christianity propaganda.⁶⁰

Hu's worries were later intensified when Nanking's A. N. Steward took up studies at the arboretum's sister institution, the Gray Herbarium, in 1928.

Chinese botanists also had occasion to feel somewhat competitive with the collectors sent out by Western institutions. Once, when Hu sent material to the arboretum for identification, curator Alfred Rehder let him know that it would be quite a while before they got to it because everyone was working on material just sent back by Joseph Rock.⁶¹ Identifications done by Western institutions were not just a favor, since this work provided an opportunity to publish new material. When Hu

---

⁵⁹ Charles Sargent, letter to E. D. Merrill, 11 July 1917 (Charles Sargent Letterbook, vol. 9; Arnold Arboretum Archive, Harvard University).
⁶⁰ Hu Hsien-su, letter to John Jack, 30 September 1926 (AACC).
⁶¹ Alfred Rehder, letter to Hu Hsien-su, 24 March 1928 (AACC).

considered his specimens were not being identified by the arboretum rapidly enough, he applied pressure by announcing his intention to send material elsewhere, for example, to W. W. Smith in Edinburgh.[62]

Accession records in the reports of the Fan Memorial Institute of Biology show exchange relations with a large number of institutions, both domestic and international. Hu's relationship with the arboretum suggests that these relations were dynamic. Since all parties had alternatives, each could seek, if not always attain, arrangements to his satisfaction. One would expect that the vitality produced by the competition was in part responsible for the rapid development of taxonomy in China.

But this vitality depended on participants' integrity in avoiding destructive competition. For example, a Chinese institution supplying duplicate sets of specimens on a fifty-fifty basis to a sponsoring Western institution could find itself in competition with its sponsor if both tried to exchange or sell identical duplicate sets to the same third-party institutions. It is to Merrill's credit that he tried to avoid this problem. After leaving the directorship of the New York Botanical Garden for Harvard in 1935, he insisted that the Botanical Garden's board of managers carefully select the places where they would exchange duplicates of Hainan Island specimens so as not to "interfere with those [exchanges] arranged by Ch'en [Huan-yung] for the remaining sets of duplicates of the same collection."[63]

The Sino-Japanese War Disrupts Research

Relations between Chinese and Western institutions vigorously expanded until the beginning of the Sino-Japanese War in 1937. The war did not bring botanical research in China to a halt—collecting in southwestern China continued—but it did stifle growth. Work went on at the Fan Memorial Institute during four-and-one-half years of the

---

[62] Hu Hsien-su, letter to Alfred Rehder, 20 March 1928 (AACC).

[63] E. D. Merrill, "Memorandum," 14 April 1936 (Memoranda from Merrill; Arnold Arboretum Archives, Harvard University). For Merrill's original assurances to Ch'en, see his letter of 24 December 1930 to Ch'en Huan-yung (Merrill Correspondence, China; New York Botanical Garden Archives).

war, "in spite of the pressure, espionage and ill-concealed hatred of the Japanese authorities."[64] In October 1940 Hu became chancellor of the new National Chung Cheng University in unoccupied Kiangsi, where he stayed until the war ended.[65] Before Japan attacked the United States, the Fan Memorial Institute was able to stay open because of its American ties through the China Foundation, a Sino-American institution. When the Japanese invaded Pearl Harbor on 7 December 1941, the institute was confiscated along with British and American institutions.[66] In April 1946 Hu returned to Peking to pick up the pieces. Fortunately, the institute's herbarium was still intact.[67]

Ch'en Huan-yung worked at the Botanical Institute at National Sun Yat-sen University in Canton until shortly after the city fell to the Japanese in October 1938. In 1947 Ch'en recounted the story of those days to Merrill.

> Bombs fell on the compounds of our Institute. . . . You suggested removal to Hongkong in readiness for instant shipment of the herbarium and library to New York, for the duration, at your expense. . . . We moved somehow. Finally Canton was completely evacuated but I slipped alone into Shameen to urge the American and British Counsels [sic] to intercede with the Japanese to spare the Experimental Gardens. The Japanese used Germans to search residences of Shameen for Chinese refugees. They came to my hiding place at midnight but I tricked the Nazis. When my mission failed I made my way by foot to Hongkong disguised as a coolie.[68]

Ch'en made another attempt to obtain sanctuary for botanical work after the capture of Hong Kong on Christmas Day, 1941. He asked the director of education of the Japanese puppet government in Kwangtung

---

[64] Hu Hsien-su, "Foreword," *Bulletin of the Fan Memorial Institute of Biology*, n.s. 1 (1943): 1.

[65] Hu Hsien-su, letter to E. D. Merrill, 17 November 1940 (AACC).

[66] Hu Hsien-su, letter to E. D. Merrill, 22 April 1940 (AACC); and "Foreword," 1.

[67] Hu Hsien-su, letter to E. D. Merrill, 20 April 1946 (AACC).

[68] Ch'en Huan-yung, letter to E. D. Merrill, 25 January 1927 (AACC).

for permission to move the botanical collections of Sun Yat-sen University back to Canton. After the Japanese were defeated, the Nationalist government charged Ch'en with "cultural collaboration" with the enemy because of his willingness to deal with the puppet government. Ch'en was eventually cleared of the charge and reinstated to his former positions.[69]

Conclusion

Two centuries of research on the flora of China established Western institutions as the centers for this study. By the time Chinese botanists took up this work in the twentieth century, the study of the flora of China was a prime and expanding research area in the West. Two historical trajectories coincided: one was the growth of a corps of talented and dedicated Chinese botanists committed to studying the flora of their country; the other was the increasing interest in China's flora by Western botanists who had a mature institutional system at their disposal. This situation made it possible for Chinese researchers to rapidly integrate their institutions into the existing system of botanical exchange while also securing the patronage of leading Western research centers. Chinese and Western botanists working both in China and abroad interacted to form a vigorous and dynamic system.

---

[69] Ch'en Huan-yung, letter to E. D. Merrill, 15 January 1947 (AACC).

The Development of Geology in Republican China, 1912-1937

Tsui-hua Yang

Geology in the West lost much of its glamor at the end of the nineteenth century, when the focus of scientific fashion shifted to quantitative theory. In Republican China geology remained at the forefront. To understand why this was so, it is necessary to look at its social circumstances.

The Geological Survey of China, inaugurated in 1912 and formally founded in 1916 in Peking, was the first and best established scientific research institution in China. Its systematic study of the country furnished geologists with a wealth of empirical data; and the studies of soil, fuel, geology, paleontology, and seismology undertaken under its auspices spurred the development of the profession. In addition, the survey's pioneering efforts can be credited with the formation of such related organizations as the Geological Society of China and the Research Institute of Geology of Academia Sinica.

The importance of the Geological Survey has long been recognized by historians and scientists. Charlotte Furth, in her study of the Chinese geologist Ting Wen-chiang, called it "the best scientific organization in Republican China"; the survey "had a legitimate position in the international learned world: its scholars were known; its journals were read; its research made a genuine contribution to knowledge of the natural history of the earth."[1] Edward C. T. Chao, a research fellow of the U.S. Geological Survey, has written that "prior to 1950, geology in China, with a history of about fifty years, was not considered to be far behind the West, unlike physics or engineering."[2]

These assessments aside, no scholarly study has treated the growth and scope of China's geological research in its social and institutional contexts. This essay will explore Chinese efforts in the Republican period to institutionalize geological research within the broad perspective of the scientific modernization of China. How were the geological associations and institutions formed? What kind of people became committed to geological research? What role was assured by the Chinese government, and how did foreign explorers, scientists, and financial support influence the development of China's geology? What was the relationship between geological research and Chinese industry, government, and society? What did Chinese geologists contribute to the study of world geology? Answers to these questions, which involve both "internal" and "external" approaches to the history of science, should help to provide an explanation for the growth of geological science in Republican China.

Geological Education and Institutions

Geological teaching in China began in 1909 at the geology department of the Imperial University of Peking, but shortly thereafter the department disbanded without producing any graduates. In the late Ch'ing, Chinese scholars and foreign collaborators such as John Fryer and John MacGowan translated Western works on mineralogy and geology. In 1872–73 Hua Heng-fang (1833–1902), an imperial mathematician who worked in the translation department of the Kiangnan Arsenal in Shanghai, translated into Chinese two important works of nineteenth-century geology—James Dana's *System of Mineralogy* and Charles Lyell's *Elements of Geology*. Hua's translations of chemical and mineral terms were often difficult for Chinese readers to understand

---

[1] Charlotte Furth, *Ting Wen-chiang: Science and China's New Culture* (Cambridge, Mass., 1970), 57.

[2] E. C. T. Chao, "Progress and Outlook of Geology," in *Science in Communist China*, ed. Sidney Gould (Washington D.C., 1960), 500.

and memorize. The author himself confessed that Western geological concepts were so difficult to grasp that his mind was haunted by dreams of the fantastic prehistoric animals Lyell described.

Despite the heroic efforts of these and other translators, their work contributed only peripherally to the development of China's geological science, chiefly because Ch'ing officials and intellectuals considered geology to be an instrument for mining, rather than an independent field of learning. Not until the Republican period did the government employ the term "geology" (*ti-chih*) and Chinese interest in earth studies evolved into a scientific discipline. Government support and foreign assistance, combined with Western-trained geologists returning from universities abroad, made it possible for professional geological studies to develop on Chinese soil.

In 1913, the Bureau of Mines under the Ministry of Industry and Commerce established the first school of geology in China—the Geological Institute (Ti-chih Yen-chiu-so). Among its first faculty members were Chang Hung-chao (1877–1971), a graduate of the University of Tokyo; Ting Wen-chiang (V. K. Ting, 1887–1936), who studied zoology and geology at the University of Glasgow; and Weng Wen-hao (W. H. Wong, 1889–1971), a doctor of geology and physics educated at the University of Louvain in Belgium. Using equipment, facilities, and dormitories borrowed from Peking University, the institute functioned as a state-funded training center for geology professionals. It closed in 1916 after producing eighteen graduates (the pool from which the Geological Survey of China would select its nucleus), when Chang and Ting argued that the government should not continue to expend its energies on pedagogy and that *universities* should assume responsibility for educating and training students of geology.

In 1918, after Ts'ai Yuan-p'ei took charge of a revitalized Peking University (Pei-ta), the geology department was restored. Hampered from the outset by inadequate facilities and a dearth of faculty, the department suffered serious curricular and pedagogical deficiencies until 1920, when two new faculty members—Amadeus W. Grabau (1870–1946) and Li Ssu-Kuang (J. S. Lee, 1889–1971)—were appointed and enrollments began to rise. Grabau, of German descent, had received his doctorate from Harvard and had been teaching at Columbia since 1901. He was a leading scientist in paleontology, stratigraphy, and sedimentology during the early decades of this century.[3] Li, a native of Huangkan, Hupei province, had studied engineering in Japan and received his doctorate in geology from the University of Birmingham (England) in 1926. Under their direction, the emphasis of instruction shifted from metallurgy and mining engineering to historical and physical geology.

When Chiang Mo-Lin reorganized Pei-ta in 1931, Ting Wen-chiang was invited to teach in the geology department. A new geological building was erected in 1935. By that time, the other four universities also supported departments of geology or earth studies. They were Chungshan (formerly Kwangtung), Chungyang (formerly Tungnan), Chungking and Tsinghua Universities. Because of its long history and its large, well-qualified faculty, Pei-ta was the most sophisticated one. According to Chang Hung-Chao, of the 264 geology students who had graduated from all Chinese universities by 1936, 188 were from Peking University.[4] The small number of geology students, who comprised 2.25 percent of all college students,[5] and the five university programs indicate that geology was by no means a popular discipline. In general, students carried out little research, but Pei-ta geology students, with better training and a strong commitment to research, had organized a society and published their papers in the department's bulletin. Pei-ta graduates had better opportunity to study abroad or to work at geological surveys and research institutions.

During the Republican period, three national or state-controlled institutions carried out geological research in China: the Geological Survey of China, the Research Institute of Geology of Academia Sinica, and the Geological Institute of the National Academy of Peiping (Peking). Headed by Ting Wen-chiang, the Geological Survey invited foreign specialists such as J. G. Andersson, a former director of the

---

[3] On Grabau, see V. K. Ting, "Biographical Note," *Bulletin of the Geological Society of China* (hereafter, *BGSC*) 10 (1931): iii–xviii; and Marshall Kay, "Grabau," in *Dictionary of Scientific Biography* (1972), Vol. 5, 486–88.

[4] Chang Hung-chao, *Chung-Kuo ti-chih-hsüeh fa-chan hsiao-shih* [A brief history of China's geology] (Shanghai, 1940), 40–41.

[5] Chinese Ministry of Education, *Erh-shih-san nien tu ch'uan-Kuo Kaoteng chiao-yü t'ung-chi* [Statistics of China's higher education in the academic year of 1934] (Nanking, 1936), 32–34.

Geological Survey of Sweden, and his assistant, F. R. Tegengren, to serve as mining advisors. Several returned students, such as Yuan Fu-li, who was educated at Columbia University, and Tung Chang, who returned from Japan, joined the survey as research fellows. During the first ten years, the survey had about 30 employees. By 1935, the number had doubled.

From its inception, the Geological Survey concentrated on area studies of mineral resources and on geological mapping, in order both to improve the native mining industry and to discover new ore deposits. This practical approach encouraged public support by serving the public interest. Mining companies such as K'ai-lan and Pei-p'iao contributed funds for the erection of a library, museum, and administration building; and several Chinese entrepreneurs, including Chin Shao-chi (Sotsu G. King) and Lin Hsing-kuei, were instrumental in helping the survey to build a fuel laboratory and to erect a seismological station in the Western Hills near Peking.[6] Such successes prompted Ting Wen-chiang to write in 1936 that "the Survey is the only government institution that has been able to get financial support from private sources in this way."[7]

From 1921–1928, the Geological Survey weathered a period of fiscal austerity. The repeated civil wars and political turmoil meant that the government could provide only erratic and generally inadequate funding, and this forced the survey to look elsewhere for financial support. In 1925 the China Foundation for the Promotion of Education and Culture (organized to manage Boxer Indemnity funds from the American remissions of 1924) began to supply substantial financial support to the survey. The China Foundation also furnished funds for a soil laboratory, which was entrusted to the survey. An examination of all grant-in-aid given by the China Foundation to eight scientific disciplines during its first ten years (1926–35) shows that the largest share went to geological science. The Geological Survey was the main organization receiving this support and would not likely have survived without it.[8]

---

[6] Geological Survey of China, *Chung-Kuo ti-chih tiao-ch'a-so kai-k'uang* [Geological Survey of China, 1916–31] (Peking, 1931), 3–5.

[7] V. K. Ting, "Modern Science in China," *Asia* 31 (1936): 131.

[8] Jen Hung-chün, "Shih-nien lai chung-chi shih-yeh te hui-ku" [A retrospect of the past ten years' work of the China Foundation], *Tung-fang tsa-chih* [Eastern miscellany] 32, no. 7 (1935): 19–25.

Government support for geological research resumed in the 1930s. The establishment of the Nanking government in 1928 brought some political unity to China after a decade of civil war. Over the next decade, the central government achieved the greatest degree of fiscal control over China since 1911, including the capacity to initiate economic construction, industrial modernization, and scientific research. In cooperation with the National Academy of Peiping, the Geological Survey became a division of the Ministry of Industry and began to develop rapidly and steadily. Leaving a branch office in Peking, the survey's headquarters eventually relocated to Nanking in 1935. By 1937, the survey contained paleontology, Cenozoic, fuel, and soil laboratories, a seismological station, a library, and a museum. It maintained several publications to report the results of its research: *The Geological Bulletin (Ti-chih hui-pao), Geological Memoirs, Paleontologia Sinica, Statement of Chinese Mining Industry, Seismological Bulletin (Ti-chen chuan-pao), Soil Bulletin (Tu-jang chuan-k'an),* and *Contributions from the Fuel Research Laboratory (Janliao yen-chiu chuan-pao).*

The institutionalization of science had been discussed by both Kuomintang (KMT, the Nationalist Party) leaders and Chinese scientists in the early years of the Republic, but it did not actually begin until 1928, when the nationalist government achieved political stability and unity in central-coastal China. Together with several prominent KMT figures, Ts'ai Yuan-p'ei engaged in education reform and advocated the establishment of a central scientific research academy. Conflicting opinion came from another influential KMT spokesman, Li Shih-tseng, who proposed building several provincial or regional academies in addition to a central one. By way of resolving the conflict, the Nationalist government decided to establish two national academies of science—the Academia Sinica in the south and the National Academy of Peiping in the north. Ts'ai and Li, respectively, were appointed presidents of these new bodies.

The Research Institute of Geology of Academia Sinica consisted of four sections: stratigraphy and paleontology, petrology and mineralogy, applied geology, and dynamic geology and geophysics. Its official publications included *Memoirs (Chi-k'an), Monographs (Chuan-K'an)* and *Contributions (Ts'ung-k'an).* With 27 employees in 1931, it was the largest among the research institutes of natural sciences in Academia Sinica. In scope it differed little from the Geological Survey, but it paid more attention to theoretical geology, such as geotectonics and the glacial study of the Quaternary. Under the direction of Li Ssu-kuang, who also taught at Pei-ta, geologists at the institute engaged in the study of structural geology with a geophysical approach.[9]

The Geological Institute of the National Academy of Peiping was in fact an affiliate organization of the Geological Survey and shared with it all facilities, equipment, and even its director, Weng Wen-hao. After 1929, the institute and the survey jointly published bulletins, memoirs, and other materials, a clear indication of their close relationship.[10] These two organizations could be considered as one and the same.

In addition to these national institutions, several smaller, provincial geological surveys (e.g., in Honan, Hunan, Kiangsi, Kwangtung and Kwangsi) and a private institute, the Western Academy of Sciences (Hsi-pu K'o-hsüeh-yüan) in Szechwan, also conducted geological studies. Since their primary mission was to provide advice to local mining companies in search of ore deposits and to perform various services for provincial governments, their research was practical in orientation. Accepting minimal financial support and technical help from the central survey, the contributions of these institutions to the geological sciences could not compete with those of the national ones.

Distinct from the private and governmental institutions, the Geological Society of China, organized in 1922, played an important role in the development of China's geological knowledge. The society was an autonomous body, with no obligation to serve either government or industry. It focused on topics of purely scholarly interest. A survey of seventeen volumes of its Bulletin, which contain 439 geological monographs, highlights this orientation: studies of paleontology, regional geology, historical geology and petrology comprise 71 percent of all monographs.[11] A comparison of this publication with the contents of the *Bulletin of the Geological Survey* suggests that the society paid less attention to applied or economic aspects of geology.

---

[9] On the Academia Sinica, see *Chung-yang yen-chiu-yüan kai-k'uang* [A survey of the Academia Sinica] (Taipei, 1957).

[10] On the National Academy of Peiping, see *Kuo-Li Pei-ping yen-chiu-yüan yüan-wu hui-pao* [Bulletin of the National Academy of Peiping] (Peking, 1930 37).

[11] Chi Jung-sen, *Chung-kuo ti-chih hsüeh-hui kai-k'uang* [A Survey of the Geological Society of China] (Chungking, 1942), 30–31.

The nucleus of the society was the workers of the Geological Survey and certain academic figures. Its membership was 71 in the first year. Fifteen years later, it increased six-fold. Foreign scientists played an active role from the very beginning. Among the twenty-six charter members of the society were three Westerners: J. G. Andersson, A. W. Grabau, and Luella Miner, chancellor of Peking Union Girls' College. Foreign fellows accounted for at least one third of all society members in 1922 and numbered sixty-three, or one half the membership, in 1926. The society published its bulletin almost entirely in English, thereby readily winning the attention of the international scientific community. It was not until 1936 that the society began to distribute another bimonthly journal, *Ti-chih lun-p'ing* (Geological Review), which was published in Chinese for the purpose of a domestic exchange of ideas. Understandably more successful in China than the Bulletin, *Ti-chih lun-p'ing* helped to popularize geological knowledge. Its appearance demonstrated the growing independence and maturity of the second generation of Chinese geologists.

Geological Research

This section will focus on three areas—economic, historical, and structural geology—leaving aside the fields of mineralogy, petrology, sedimentation, seismology, and soil studies, which played a less significant role in the development of geology in China.

In a broad sense, economic geology is concerned with materials of practical utility to humanity. More narrowly, it includes the study of ore deposits, petroleum geology, and the geology of non-metallic deposits, and this involves the application of several disciplines within the geological sciences. Exploration for deposits of metallic ores, for example, is guided by mineralogy, petrology, and stratigraphy, often coupled with applied aspects of geophysics and geochemistry. While the "scientific" investigation of ore deposits, which examines and explains the composition, shape, formation and genesis of metal concentrations within the earth's crust, is not entirely determined by economic considerations, the practice of economic geology helps support the exploration for and extraction of mineral resources.

The development of geological science in China has been closely related to practical applications. Translations of Western earth studies by late Ch'ing scholars were intended to aid the exploration of Chinese natural wealth and the modernization of its mining industries. Chinese geologists of the early twentieth century surveyed and studied mineral resources for national construction and industrialization. Consequently, many Chinese regarded geology almost exclusively as mineral exploration or prospecting, and geologists as mining technicians or engineers. Weng Wen-hao commented on this "misleading" concept of geology and proclaimed that "theoretical" and "applied" aspects of geology were equally important. Concerned with the interdependence of theory and application, Weng suggested that Chinese geologists pay equal attention to both.[12]

Nevertheless, most Chinese geologists took a lively interest in practical mining development. The members of the Geological Survey of China, in particular, emphasized the close relationship between geology and mining, prompted by the belief that China's mineral resources had been either unstudied, misinterpreted, or exaggerated by early foreign explorers.[13] In order to compete with foreign mining interests in China and to gain a clear idea of the present condition of mineral resources, Chinese geologists helped the government search for new mineral lodes, develop small Chinese-owned mines, collect mining data, and make estimates and statistical studies of mineral reserves and production.

These investigations were closely connected to the expansion of the railways. The Ministry of Communication and several railway administrations entrusted the Geological Survey with the exploration of mining fields along railroad lines. The results of these missions were published in the Survey's *Bulletin,* which illustrated pioneering maps and surveys as well as collections of specimens. Weng Wen-hao further organized the existing body of knowledge, publishing *Mineral Resources in China* in 1919, and compiling and editing the Survey's first *General Statement of the Mining Industry* in 1921. The latter was continually revised and had gone through five editions by 1935.

Geologists often had a personal as well as scientific interest in the mining industry. Ting Wen-Chiang, for example, resigned the directorship of the Geological Survey in 1921 to become manager of the Pei-

---

[12] Weng Wen-hao, "Li-lun te ti-chih-hsüeh yü shih-yün te ti-chih-hsüeh" [Theoretical geology and applied geology], *BGSC* 4 (1925): 185–92.

[13] W. H. Wong, "Chinese Geology," in *Symposium on Chinese Culture,* ed. Sophia H. Zen (Shanghai, 1931), 208–9.

p'iao coal mine in Jehol. Under his management, the Pei-p'iao mine by the late 1920s had become one of the most productive and profitable small Chinese mines. Ting subsequently published three pamphlets on contemporary mining operation: *Mining in China during the Past Fifty Years, Outline History of Chinese Officially Managed Mines,* and *Materials for a History of Foreign Capitalized Mining.*

Geologists clearly had direct impact on the practical development of the mining industry. Yet how significantly did the study of mining and mineral resources promote the science of geology? The answer lies in the relationship between the "practical" and "scientific" aspects of economic geology and is best illustrated by geologists' studies of coal and iron deposits.

Coal deposits are widely distributed throughout China. According to Juan Wei-chou's 1945 study, 83 percent of China's coal was located in the provinces of Shansi and Shensi.[14] These deposits were situated close to sources of iron ores in North China and most were suitable for making metallurgical coke. These factors were encouraging for the future of China's industrialization and spurred the study of coal geology. Compared with coal resources, however, iron reserves in China were less promising for building a steel or other heavy industry. High-grade iron deposits were located in North China (except in Hopei), the Yangtze valley, and the southwestern region, but the mines were small and widely scattered. The successful manufacture of iron depended upon the cost of mining, processing, transporting, and marketing the product. These costs in turn depended upon the location of the deposits.

Chinese geologists concentrated their explorations for coal and iron resources in the North China plain and Yangtze valley. By the end of the 1930s, the Survey's *Bulletin* had published more than forty detailed reports on coal fields. These included valuable maps (mostly on the scale of 1:50,000 or 1:100,000), descriptions of coal-bearing rock units, studies of the stratigraphical and structural relationships of coal fields, and research on a variety of problems related to fuels. As for the exploration of iron ores, the survey proudly announced in 1931 that the total reserves of iron deposits discovered by its members amounted to 140 million tons out of the grand total of one billion tons actually

---

[14] On the geographical distribution, geological occurrence, and production of Chinese coal, see Juan Wei-chou, "Mineral Resources in China" (Ph.D. diss., The University of Chicago, 1945), 12–23.

known.[15] The early results of their labours were summarized in Tegengren's *Iron Ores and Iron Industry of China*, published from 1921 to 1923. From a practical perspective, this work not only described the iron resources of China, but gave a historical account of the development, production, and inputs and outputs of both "traditional" and "modern" iron industries. Tegengren's approach revealed a weakness that characterized all early geologists' studies of mineral resources— namely, that the geological environment, the precise classification of individual ores, and the mineralogical or petrological analyses were either undervalued or ignored entirely.

Students returning from universities in the United States brought to China Lindgren's theory of mineral deposits and thereby laid the foundation for advanced geological study. The leading contributor to the science of ore deposition in the first half of the twentieth century, Waldemar Lindgren (1860–1939), was recognized as one of the world's foremost economic geologists. His textbook, *Mineral Deposits*, published in 1913, is still valued for time-tested genetic theories, and his classification of mineral deposits has undergone only minor revisions. A founder of the journals *Economic Geology* and the *Annotated Bibliography of Economic Geology*, he taught in and then headed the department of geology at the Massachusetts Institute of Technology from 1912 until 1933.[16]

At MIT, Lindgren helped train such distinguished Chinese students of geology as Wang Chu-chuan, Chang Keng, Meng Hsien-min, and Cheng Hou-huai. After returning to China, Wang joined the Geological Survey, while Chang and Meng signed on with the Research Institute of Geology of Academia Sinica. On Lindgren's recommendation, Cheng was employed by the U.S. Geological Survey, and the two remained in close contact after Cheng's return to China, where he taught economic

---

[15] Geological Survey of China, *Chung-kuo ti-chih tiao-ch'a-so kai-k'uang*, 12–13.

[16] On Lindgren, see L. C. Graton, "Life and Scientific Work of Waldemar Lindgren," in *Ore Deposits of the Western States, Lindgren Volume* (New York, 1933), xiii–xxxiii; M. J. Buerger, "Memorial of Waldemar Lindgren," *American Mineralogist* 25 (1940): 184–88; Huang I, "Lin-ko-lan hsien-sheng chih sheng p'ing chih-hsüeh chi ch'i hsüeh-shu kung-hsien" [Mr. Lindgren's life, scholarship, and contributions], *Ti-chih lun-p'ing* [Geological review] 4 (1940): 441–48.

geology at Chungyang University. Thus, through Cheng, Lindgren's stress on igneous processes in ore formation and the application of the petrographic microscope to the study of ores and their constituent minerals had great impact on the study of mineral deposits in China.

Using Lindgren's theories, Chinese geologists conducted field studies in South China, making detailed and precise reports on iron ores and emphasizing patterns of formation and distribution. In 1935 Hsieh Chia-jung (1898–1966) and others published *Geology of the Iron Deposits in the Lower Yangtze Region*. In this work, petrographical and microscopic methods were used to examine the physical properties and chemical constituents of rocks and ores, their types and genesis. These enabled the crystallographic measurement and chemical analysis of minerals such as apatite, garnet, and adularia, thereby revealing important relationships between the origin of deposits and mineral composition of the ores. By considering the relation between the geological age and the type of deposits, Hsieh drew an outline of the important iron-forming epochs in China. This work not only provided detailed information on the Yangtze iron ores from a local point of view, but at the same time made a contribution to the study of ore deposits in genetic relation to igneous intrusion.

Meanwhile, the Sinyuan Fuels Research Laboratory of the Geological Survey began the chemical and mineralogical study of mineral fuels, particularly coal and synthetic fuel oil. Headed by Hsieh Chia-jung, the Sinyuan lab produced many reports on chemical and physical studies of coal and on microscopic analysis of Chinese coal. Hsieh himself published a number of microscopical and petrographical studies of Chinese coal and invented some new techniques in coal petrography. Combined with coal chemistry, the study of coal petrography contributed to the precise classification of coals and the study of their genesis and formation, which, after 1949, would become an important area of coal geology.

In addition to the study of individual ores, the broader problem of deposit formation in time and space received special attention. Mineral deposits are not uniformly distributed in the earth's crust, nor do they all form at the same time. Regions in which conditions are favorable to the concentration of useful minerals are termed "metallogenetic provinces" and contain broadly similar types of deposits. The concept of metallogenetics originated in the late nineteenth century and was enlarged by Lindgren's coining of the term "metallogenetic epoch" to characterize those times in earth history when, in one or more large regions, conditions for ore deposition were especially propitious.

Weng Wen-hao introduced these space and time concepts of ore formation into Chinese geology. Emphasizing the magmatic source of ore deposits, Weng paid much attention to the problem of "regional zoning." He traced the distribution of metallic ores in southern China, dividing the region into four zones: tin-tungsten-bismuth, copper-lead-zinc, antimony, and mercury.[17] Although this preliminary scheme provoked criticism, Weng had established a classical model—his examination of the rules for the distribution of metaliferous deposits and the ages of their formations laid the foundation for the study of metallogenetic provinces and epochs, and his discussions of the positional relations in the metallogenetic series continued to command considerable attention in the 1970s.

In contrast to Weng's concern with geographic and stratigraphic distribution of metallic deposits, some young geologists approached the temporal and spatial relation of ore deposits from the standpoint of structural connections. Wang Ch'ung-yu examined various forces in the dynamics of mountain formation and concluded that marginal mountain chains were the most favorable regions for mineralization.[18] Hsieh Chia-jung studied the relation of crust movement and igneous activity to metallic deposition. His discussions of tectonic features were especially pertinent to the temporal dimension of ore formation. He regarded the late Mesozoic and early Tertiary periods as the major metallogenetic epoch in China.[19]

These generalizations about space-time-genesis relations brought order, consistency, and meaning to the study of ore deposition. Sometimes based on narrow experience, however, these observations were not always highly valued. Generalizations unsupported by independent local

---

[17] Weng Wen-hao, "Chung-kuo K'uang-chan chu-yu lun" [Discussions on the metallogenetic provinces of China], *Ti-chih hui-pao* [Bulletin of the Geological Survey of China] 2 (1920): 9–24.

[18] Wang Ch'ung-yu, "The Relation of Tectonics to Ore Deposits," *BGSC* 3 (1924): 169–82; "The Relation of Ocean Deeps and Geosynclines to Ore Deposits," *BGSC* 5 (1926): 25–35.

[19] Hsieh Chia-jung, "On the Late Mesozoic–Early Tertiary Orogenesis and Vulkanism and Their Relation to the Metallic Deposits in China," *BGSC* 15 (1936): 61–74.

tests or dependable and sufficient facts are usually bad risks in ore exploration. Wang Ch'ung-yu's hypotheses on large tectonic consideration, for example, eventually proved premature.

One objective of geological science is to work out the full history of the earth and its animal and plant inhabitants. There can be no doubt that Chinese interest in historical geology was largely inspired by A. W. Grabau. Grabau's paleontological treatises, stratigraphic studies, and textbooks, written before he came to China, had had profound influence on early twentieth-century historical geology. During his twenty-six years in China, in addition to giving lectures at Peking University, Grabau took charge of the paleontology lab of the Geological Survey and produced more than one hundred publications in various professional journals and magazines. Under his guidance, a dozen Chinese paleontologists and another dozen geologists began to engage in research on the subject.

The economic, as well as methodological and philosophical, implications of historical geology prompted Chinese interest in this science. First, historical geology has practical value. Stratified rocks include marketable resources of many kinds—potash, sodium and other salts, limestone, and numerous metalliferous deposits. An industrial society runs on fossil fuels—coal, oil, and natural gas—which occur almost exclusively in stratified rocks. Historical geology has thus been applied extensively to petroleum exploration, which relies heavily on the skills of the paleontologist, the sedimentologist, and the stratigrapher.

Second, historical geology employs research methods similar to those used in the study of history. Furth, in her study of Ting Wen-chiang, remarked that historical science makes use of empirical methods that "had some parallels with the concerns and methods of Chinese traditional scholarship." She explained this phenomenon as follows:

> The first generation of Chinese students of Western science were in general attracted to empirical and historical disciplines: medicine, geology, biology, and archaeology. Relatively few of them became physicists or chemists. Both in practice and in account of how that practice operated, the natural history sciences of the nineteenth century were more immediately accessible to a Chinese whose forebears had considered historical studies the model for all rational inquiry, and who was not

accustomed to distinguish between theoretical concepts in science and speculative thought in general.[20]

Some Chinese geologists of the time also took note of the relationship between geology and history. Yang Chung-chien (1897–1979), a vertebrate paleontologist, remarked in 1941 that the methods of historical geology, which focus primarily on the collection of data, are "the same" as historical studies, though the materials and subjects of these two disciplines are different. While historians are concerned with human beings and their written documents and records, geologists study rocks and fossils in order to understand the evolution of the earth and life on it.[21]

This view was shared not only by Chinese geologists, but by a group of Japanese field researchers, the so-called "grass-roots geologists." Nakayama Shigeru's study of the Chidanken (Society for Corporate Research in Earth Science) sheds light on this issue. The Chidanken, founded in 1947, was not merely a scientific society, but a crusading body whose purpose was to propagate its ideology and methodology. Its able organizer, Ijiri Shōji, served as its head for two decades after the end of World War II. In Ijiri's view, geology is closer to the historical sciences than the exact sciences of physics and chemistry. Inasmuch as the goal of geology was to describe the history of the earth, a historical approach should be adopted as its research methodology. The method advocated by Ijiri was concentration on field work. Unlike physical occurrences, historical events have their own individual characteristics in space and time, and they must be pursued through the accumulation of data derived from field investigations in each locality.[22]

Questions, however, remained. To what extent could the geologist learn from historical methodology? Did the historical approach help Chinese geologists to resolve geological problems? An examination of

---

[20] Furth, *Ting Wen-chiang*, 56.
[21] Yang Chung-chien, "Ti-chih-hsüeh yü shih-hsüeh" [Geology and history], *Wen-shih tsa-chih* [Magazine for literature and history] 1, no. 5 (1941): 1–4.
[22] Nakayama Shigeru, "Grass-Roots Geology," in *Science and Society in Modern Japan*, ed. Nakayama Shigeru (Cambridge, Mass., 1974), 270–89.

what they accomplished in the study of historical geology may help us to answer these questions.

As chief paleontologist of the Geological Survey of China, Grabau took the lead in organizing research and the exchange of ideas. His medium was the journal *Paleontologia Sinica,* which was issued in four series: series A was devoted to fossil plants, series B to fossil invertebrates, series C to fossil vertebrates, and series D to ancient man. By 1937, 114 fascicles of the journal had been published, 8 in series A, 47 in series B, 46 in series C, and 13 in series D.

Paleobotany, the study of fossil plants and vegetation, combines a knowledge of both geology and botany. Several Chinese geologists published papers describing fossil plants, but these were the by-products of geological investigation, rather than of research undertaken specifically on this topic. The only established paleobotanist of early twentieth century China was Ssu (Sze) Hsing-chien. A graduate of Peking University, Ssu recieved a China Foundation research fellowship to study paleobotany at the University of Berlin from 1931 to 1933. Returning to China with a doctorate in the subject, he taught and worked at several universities and geological institutes, producing many monographs on paleobotany. On the whole, however, works in this field were fewer and more restricted than those in vertebrate and invertebrate studies.

The reason so few Chinese students studied paleobotany was that the field was too specialized and the training too restricted. Because it requires a detailed knowledge of plant morphology, which most paleontologists do not possess, paleobotany has come to be more a subdiscipline of botany than an adjunct of general paleontology. In China, as Ting Wen-chiang noted in 1924, the geology department of Peking University did not offer "serious" courses in biological studies, which made it "almost impossible for students to understand the fundamental principles of historical geology."[23] Certain biologists instead supplied recruits for the study of fossil plants.

While paleobotanic study remained an underdeveloped field in Republican China, Chinese paleontologists took an active part in invertebrate studies. For example, Huang Chi-ch'ing and Tien Ch'i-chün worked on brachiopods; Li Ssu-kuang and Chen Shu on fusulinidaes;

---

[23] "Presidential Address: The Training of a Geologist for Working in China," *BGSC* 3 (1924): 10.

Sun Yün-chu and Yin Tsan-hsün on graptolites; and Chang Hsi-chih, Yu Chien-chang, and Yüeh Sen-hsün on corals. By 1937 thirty-four Chinese geologists, of whom twelve were professional paleontologists, had produced 173 monographs on invertebrate studies.[24]

The studies of Chinese vertebrate paleontology published in both the West and China were primarily the work of foreigners rather than native Chinese. The Central Asiatic Expedition of the American Museum of Natural History collected mammal and Mesozoic dinosaur fossils from Mongolia; the Northwestern Scientific Expedition brought Tertiary mammals from Chinghai to Sweden; and Andersson collected a number of vertebrate fossils in Shansi and Honan and sent them to Uppsala University and to Professor Max Schlosser of Munich University. The most noted Chinese vertebrate paleontologist was Yang Chung-chien, who graduated from Peking University and studied at Munich University under Schlosser. In 1927, Yang published *Fossil Rodents of North China* in German; it was the first monograph on vertebrate fossils produced by a Chinese. In the next decade, he taught at Peking University, worked for the Geological Survey, and produced many publications on the vertebrates. Nevertheless, the works in this field could not, on the whole, compare in either quality or scope with those in invertebrate studies.

The high cost of the field and laboratory work associated with vertebrate paleontology was one reason few Chinese engaged in its study. Vertebrate fossils are generally larger and rarer than invertebrate fossils. Consequently, the collection and preparation of skeletons demand great financial support and manpower. Before the China Medical Board of the Rockefeller Foundation provided funding for a Cenozoic lab, no institution in China could afford to conduct large-scale investigations, and no qualified laboratory worker had the means to analyze vertebrate specimens.

The discovery of Peking Man gave great impetus to the study of vertebrate paleontology, especially the study of ancient man. J. G. Andersson first discovered a deposit rich in vertebrate remains at Choukoutien, a partially destroyed limestone cave near Peking. His assistant, Otto Zdansky, then found two fossilized hominid teeth. In

---

[24] Juan Wei-chou, "Ti-chih-hsüeh" [Geology], in *Chung-hua min-kuo k'o-hsüeh chih* [Monographs on sciences of the Republic of China], ed. Li Hsi-mo, 3 vols. (Taipei, 1955), 2: 15.

1926, the Geological Survey, in cooperation with the Department of Anatomy of the Peking Union Medical College, organized a research program for large-scale excavation at Choukoutien. Three years later, the Rockefeller Foundation made a grant to the Survey to establish the Cenozoic Research Laboratory, which carried on field work at Choukoutien and investigated general Cenozoic geology and paleontology elsewhere in China.

Headed by Davidson Black (1884–1934), a Canadian anatomist and anthropologist, many foreign scientists participated in field work at Choukoutien, but the principal field coordinators were Yang Chung-chien and P'ei Wen-chung (1903–1982). P'ei joined the excavation upon graduation from Peking University in 1928. The following year, he unearthed the first Peking Man skull. P'ei's work on Pleistocene paleontology, human fossils, and paleolithic culture earned him the reputation of foremost scientist in the field of fossil men in China. The primary focus of the Choukoutien excavation shifted several times. Before 1927, it was centered around paleontology, with the goal of discovering vertebrate fossil remains; from 1927 to 1930, efforts were largely devoted to anthropology and the unearthing of human fossils; and after 1930, Choukoutien became a diverse project with special emphasis on archaeology.

Within four years of his arrival in China, Grabau published the first volume of *Stratigraphy in China* (1924), dealing with Paleozoic and older rock formations. The second volume, which appeared in 1928, was confined to the Mesozoic. This work served as a compendium for the use of students and professionals in the field of Chinese geology.

Grabau formulated several working hypotheses and principles which would guide subsequent studies. The most noteworthy of these was his concept of geosyncline—a long but relatively narrow linear trough that subsided under the pressure of sedimentary and volcanic rock, accumulated over millions of years. Geosynclines were represented by the great plain of northern India, the Persian Gulf, and the great plain of China. Most of the older geosynclines had been converted into mountain ranges by crushing and folding of the strata, of which the Himalayas were the most dramatic example on the Asiatic continent.[25] From

---

[25] The idea of geosyncline was first proposed by James Hall in 1959. For the evolution of the concept and its importance in stratigraphic study, see A. W. Grabau, "Migration of Geosynclines," *BGSC* 3 (1924): 208–349.

mountain ranges and their preserved roots, the Chinese tried to find paleontological and stratigraphical evidence to construct the paleontologic history of China. Their contribution to stratigraphic knowledge has been discussed by Chang Hung-chao in his *Brief History of China's Geology*.[26]

In addition, Grabau introduced to the Chinese the concept of a paleozoic "pangaea," the theory that the continents once formed a single mass, which broke apart to form the present configuration. Grabau was an early proponent of Wegener's theory of continental drift, although Grabau had different ideas on the origin of mountain movements. In his time, only a heretic could question the permanence of continental areas and ocean basins. Grabau's views of pangaea and the origin of geosynclines led to his pulsation theory, which attributed the changing distribution of lands and seas to great rhythmic advances and regressions of the seas, which were in turn dependent on restriction and expansion of the capacities of the ocean basins: the so-called "eustatic control." Grabau departed from convention by dividing the Paleozoic series into sedimentary layers, which reflected the system of pulsation.[27]

While the ingenious concepts involved in Grabau's pulsation theory and separation of pangaea encouraged imaginative syntheses of geological evidence, authorities have never agreed on the broad picture, because Grabau's alterations of established geological classification were based on historical accidents rather than on natural principles. Geologists in the West, as well as in China, continued to use the established system. Grabau's works and methodology remained influential in many subdivisions of historical geology, but his theories and concepts on the nature of earth movement had little impact in China, because few stratigraphers had Grabau's grasp of world geology or his interest in its collation. Grabau's hypotheses nevertheless paved the way for later generations of Chinese geologists to examine and accept the theory of plate tectonics.

---

[26] Chang Hung-chao, *Chung-Kuo ti-chih-hsüeh fa-chan hsiao-shih*, 44–72.

[27] For detail, see Grabau's five-volume *Paleozoic Formations in the Light of Pulsation Theory* (Peking, 1936–38); "The Significance of the Interpulsation Periods in Chinese Stratigraphy," *BGSC* 18 (1938): 115–20.

By and large, historical geology of this period in China concentrated on the description of new discoveries, the division of biotic sequences and stratigraphic correlation. Geologists lagged behind world standards in many respects, particularly in research on systematic biology and on the evolution of morphological function and organic community, and in the application of new techniques and new methods.

The inquiries of historical geology in China stressed classification and induction. Chinese geologists produced some research along standard lines, such as increasing the precision of stratigraphical correlations and paleogeographical maps, but placed priority on field work. Such work had at least as much worth as the endless excavation of local historical sites.

In evaluating the philosophy of Ijiri and the Chidanken, Nakayama Shigeru doubts that geology should pattern itself after the historical sciences. In his view, geology is much closer to the physical sciences, the only difference being its concern with time sequences, with which historical sciences must also be concerned. Nakayama denigrated the historicism in Ijiri's philosophy of science and the Chidanken's official methodology by saying that "it may not be too profitable for the geologist to draw his methodology stimulus from the established historical disciplines and thus intentionally reject the intrusion of physico-chemical methods." He concludes that "unless it is liberated from the limitation of the framework of historical science and is open to enrichment by whatever is useful, Ijiri's philosophy of science will itself turn out to be a historically outmoded fossil."[28]

Fortunately, some Chinese geologists recognized the significance of research along pluralistic lines and proposed that earth sciences no longer remain aloof from the general trend toward reductionism that invaded all fields of natural science. Dissatisfied with the prevailing empirical approach, Li Ssu-kuang called for new methods, acknowledging the need for fundamental theories to interpret geological facts and remarking that "the old naturalists' stand" in handling geological matters was no longer adequate. From his point of view, the most important issues of "modern" geology were "the hypotheses or problems concerning orogenesis [mountain movement] and the questions of mobility of continental masses each as a whole or as integral parts of the

---

[28] Nakayama, "Grass-Roots Geology."

crust of the earth."[29] These questions and problems belonged to the domain of structural geology or geotectonics.

Although early foreign explorers had established a skeletal knowledge of the basic geological structure of China, much of the detail remained inaccessible. In the first two decades of the twentieth century, most of the Chinese geologists active in field work were stationed in the north. The simple stratigraphy and tectonic relations of this region suggested that China had a stable continental structure. At the end of the 1920s, when Academia Sinica's Research Institute of Geology was founded in Shanghai, geologists began to explore the south. There they found numerous fossiliferous marine formations, interformational breaks, unconformities of violent nature unknown in the north, and other complex structural relations. The vast area of China, covering much of the continent and bordering the world's oldest ocean, provided abundant materials for tectonic study.

Reviewing the leading hypotheses of tectonic study, Li Ssu-kuang argued that "geosynclines and mountain ranges are but a particular phase—a minor phase—of continental structure; or in other words, orogenic movement is but a critical manifestation of a more widespread movement of the continents."[30] In an attempt to see what geological processes and forces are involved in the secular changes of grand tectonic movements, Li led his students and colleagues to investigate the nature of the rotational forces on the surface of the globe. They approached these problems by analyzing complex stress patterns on structures of rock formations, rather than surveying the detail of tectonic relations. As Li explained:

> Unless and until we have had a complete survey of the distribution of forces or the arrangement of a system of stresses in the rocks involved we can hardly draw any safe conclusion as to the nature of the earth-movement. It seems advisable to guard against these pitfalls in deducing the mechanical significance of folds and faults. Thus, it is evident that the analysis of complex stresses as deduced

---

[29] J. S. Lee, "Reflection on Twenty Years' Experience," *BGSC* 22 (1942): 24–27.

[30] J. S. Lee, "The Fundamental Cause of Evolution of the Earth's Surface Features," *BGSC* 5 (1926): 209–62.

from various mechanical structures of rocks would be an essential step toward the understanding of the dynamic significance of tectonic features at large.[31]

Li emphasized the importance of deducing the vectors of the principal stress field responsible for producing the linear fold and fault patterns that cover much of China. He sorted the salient structural features of China into three dominant groups: the Cathaysian and neo-Cathaysian trends (the former dating from the Paleozoic and generally running north-east, the latter probably late Mesozoic and running north-north-east); the east-west fold-zones (probably of deep-seated nature); and the various shear-forms.

From laboratory experiments with plastoelastic materials, Li was able to infer that these features in the earth's crust, particularly the shear-forms, resulted from the horizontal component of the centrifugal force arising from the spin of the earth or, more probably, from an increase in the earth's rotational speed. This conclusion was based on experiments that subjected clay and paraffin models to the effects of rotational and tilting forces. The Research Institute of Geology of Academia Sinica installed a torsion balance, Drehwage, or Variometer, with the intention of using it for the determination of gravity in different regions, in order to show the isostatic condition of various structural elements in China. Research fellows of the Institute also made a series of experiments that tested the elasticity of rocks, by measuring the vibrations of rock-slides caused by passage of electrical current. These experiments, familiar to physicists, brought geologists to a subject closely connected to geophysics and physiographical research.

Li's attempts to classify structural features and mountain forms were described in his book *The Geology of China,* published in London in 1939. His tectonic theories developed into a school of "geomechanics"; his emphasis on geophysics guided the future direction of geological research in China. After 1949, Li became a prominent figure in the People's Republic of China and had a great influence on both science policy and geological research. The discovery of oil in continental facies basins in China and the belief that earthquakes can be forecast are directly attributable to his insight. One striking proof of Li's influence is the 1:4,000,000 tectonics map of China published in 1976, which

---

[31] Lee, "Reflection on Twenty Years' Experience," 33.

portrays the various structural elements Li proposed. The Institute of Geomechanics of the Chinese Academy of Geological Sciences (under direct control of the State Bureau of Geology) has successfully applied geomechanical analysis to the exploration of mineral deposits.

Conclusion

This study of the institutional and intellectual development of geological sciences in China ends with the year 1937, when the Japanese invasion forced the Nationalist government, as well as the majority of scientific institutions and scientists, to retreat from the coast to the hinterland of China. Scientific development in general, and geological research in particular, at this point embarked on a new phase of development. The National Resources Commission and the National Economic Council, rather than the Geological Survey, played the leading role during wartime in exploring for mineral resources, conducting engineering projects, and developing the frontier.

After 1949, geologists became concerned with the new direction and new mission of geology. They began to define geology as a science originating from the process of production and practice and determined by the demands of social development and productive forces. Geological science was to become a "national" and "Chinese" science, as well as a science of the people. When Joseph Needham revisited China in the early 1950s, he was astonished that geological sciences were seen as vital for the development of the country. The Geological Survey had been given the status and power of a ministry. Needham concluded that "there can be no other nation in which the sciences of the earth are so highly regarded."[32]

Regarding the reasons for early prominence and later sophistication of geology in Republican China, this study has led to several general conclusions. First, geological science has obvious economic applications. Since the latter part of the nineteenth century, when China was threatened by the Western powers, the applied aspects of geology attracted much interest. Western knowledge of geology was introduced to China as a means to discover the natural wealth of the country and

---

[32] Joseph Needham, "Chinese Science Revisited," *Nature* 171, no. 4345 (1953): 239.

to enhance China's "wealth and strength." From then on, the Chinese government paid attention to geological work as one of the many necessary adjuncts to industrialization. Geologists were given the nationalistic duty of exploring the vast country, and studies in geology were given top priority as the vanguard of the national program of industrialization.

Second, geological science is "local" or "regional" in nature. Several Chinese scientists have discussed the question of universalism versus localism of scientific knowledge. Jen Hung-chün, for example, divided the sciences into two groups according to the material with which they are concerned: those of a universal nature, such as mathematics, astronomy, physics; and those having a more local, or restricted, character, such as botany, zoology, and geology. From Jen's point of view, scientists have a double duty—to make full use of local material for scientific treatment and to contribute to the advancement of science in general. Unless they can perform the first duty well, "they are likely to be deficient in the second duty." Jen therefore concluded that it was "natural" that "scientists in China have in recent years devoted more time and energy to the development of geology, zoology, botany, paleontology and archaeology than to physics and chemistry."[33] Yang Chung-chien also regarded geology as a "local" science, belonging to the natural, rather than exact, sciences.[34] While the principles of geological sciences are applicable to all terrestrial areas, each major unit of the continental surface displays its own peculiar features. The territory of China itself, covering more than one tenth of the earth's land surface, provided the Chinese with a large living laboratory.

Third, geology differs from other scientific fields in that it is not primarily a laboratory discipline, but a science which reconstructs the past from relics unearthed in the present. As a historical science, geology offers a greater variety of theoretical interpretations than is the case in the experimental sciences; factors of size and time make experimental study of many geological processes difficult. The special methods

---

[33] H. C. Zen, "Science: Its Introduction and Development in China," in *Symposium on Chinese Culture*, ed. Zen, 171–72.

[34] Yang Chung-chien, "Lun yen-chiu yü ti-fang-hsin k'o-hsüeh chih chi-pan kung-tso" [Discussion of the basic tasks for scientific research of local character], *K'o-hsüeh* [Science] 18 (1934): 5–11.

of geological research, particularly field work, distinguish this discipline from the experimental and physico-mathematical sciences.

The localistic character of geology has sometimes resulted in an oversimplified understanding of its nature. Daniel Kevles, a historian of physics, compared physics with geology, saying that "geological theorizing did not require the analytic training and mathematical mastery necessary in physics," and that the principal task of geology was "the elucidation of existing theories of evolution through observation or experiment."[35] Yang Chung-chien confessed that he chose geology as his major at Peking University chiefly because his family forced him to study it but also because he was a poor student in mathematics and not very interested in either chemistry or physics.[36]

Finally, geology is an analytical and complex science; it is not free from physico-chemical reductionism, even though it relies more on descriptive and taxonomic methods than do the experimental sciences. With the development of more powerful tools for producing, controlling, and registering pressure on rock specimens, structural geology has increasingly become a laboratory science. Attempts to apply the tools of physical mechanics to structural geology are difficult because of the irregular nature of most rock units and because of uncertainty as to the forces that cause rocks to deform. However, Chinese interest in the causal processes operative on the earth did gradually develop into an indigenous experimental and theoretical study of dynamic geology and geotectonics.

---

[35] Daniel Kevles, *The Physicists* (New York, 1978), 37.
[36] Yang Chung-chien, "Li Ssu-kuang lao-shih hui-i lu" [A memoir for Master Li Ssu-kuang], in *Li Ssu-kuang chi-nien wen-chi* [Memorial collection for Si Ssu-kuang] (Peking, 1981), 3.

Comment on the Historical Essays

Ka-che Yip

The preceding papers deal with the vast and relatively unexplored subject of the development of sciences in Republican China. They not only provide important details about how and why the sciences developed but also help to define some of the areas for investigation. My comments will focus on a few of the issues they raised.

The developments discussed in these papers took place at a time when the traditional orthodoxy had been undermined and a new orthodoxy had not been established. Thus, study of these developments will offer insights into what Chinese scientists had accomplished in an environment relatively free from ideological constraints; and their accomplishments were many. But were the scientists completely free from traditional intellectual influences? To put it another way, are there any particularities of Chinese culture that might have influenced Chinese scientists' interest in taxonomy and taxonomic methods? This is, in fact, one of the questions raised specifically by the papers. Tsui-hua Yang's observations on geology suggest at least a partial answer, that characteristics of taxonomy and taxonomic methods attracted the Chinese because they had long been interested in historical studies. Can we perhaps go further and find another possible clue in the classical Chinese philosophical tendency to seek order by dividing the human and physical worlds into a number of fixed categories? As Jerome Ch'en has noted, "the Chinese mind was interested in synthesis rather than analysis, in similarities rather than dissimilarities. It was happier when arranging things and ideas . . . in patterns than when developing theories to explain the patterns."[1] Further exploration of these and other cultural traits may uncover more evidence of the importance of traditional influences on the development of sciences in Republican China.

The taxonomic approach seemed to be particularly appropriate to the initial development of such "local" or "particular" sciences as geology, zoology, botany, and archaeology. Interest in these "local" sciences may indeed have been prompted by nationalistic or nativistic concerns, as the papers have suggested. But were these concerns crucial in determining a Chinese student's decision in selecting these fields for study? Laurence Schneider and William Haas offer two interesting perspectives on this important question. Schneider suggests that the reason naturalists came to outnumber experimental biologists, and geneticists in particular, may lie in Chinese intellectuals' insistence that the priority of Western science in China should be the study of nature as it is manifested in China, in spite of the fact that the mainstream of Western biology was increasingly characterized by the experimental, reductionist mode. On the other hand, Haas's study of the development of botany in China shows that research on the flora of China was one of the prime areas of study in systematic botany during the early part of this century. This interest among Western botanists stimulated Chinese botany students to seek graduate training in the study of China's flora in many Western centers. The development of this science was certainly made easier by the convergence of these students' interests with the mainstream in the West. But what if there had been no convergence? Would the Chinese students have turned to other areas of study? How firm was their commitment to "local" sciences?

There is, therefore, the broader question of the significance of local sciences. To underscore the importance of local sciences to Chinese scientists, Tsui-hua Yang quotes from Jen Hung-chün's article in the 1931 *Symposium in Chinese Culture,* in which Jen wrote that men of science in every nation have a double duty to perform: making full use of local materials for scientific treatment and contributing to the advance of science in general. But how much emphasis should Chinese scientists place on these respective duties? It is interesting to note that

---

[1] Jerome Ch'en, *China and the West: Society and Culture, 1815–1937* (Bloomington, Indiana, 1979), 176.

Peter Buck, in his book *American Science and Modern China*, quotes the same passage by Jen to show the transformation of Chinese scientists' view of local sciences and their significance. Buck argues that when the Science Society opened its Biological Research Laboratory in 1922, the intention of the founders was to explore questions related to specific Chinese conditions, in order that the nation's "desperate weaknesses" could be more readily cured. By 1931, however, these same scientists were maintaining that since Chinese fauna and flora were previously unknown to world science, their work would "strike something very important which would open up a new chapter in and make a prominent contribution to" their chosen fields of study.[2] In this light Jen Hung-chün seems to have been concerned primarily with his colleagues' "responsibility" to the cause of international scientific progress. Indeed, all the papers show that for most Chinese scientists, their points of reference were scientific communities abroad. They published in respected international journals and exchanged information with scientists in other countries. As James Reardon-Anderson has observed, "the twig of science in the Republican era was bent more to meet the standards of the international science community."[3]

In pursuing this question of nationalism in science, we must also bear in mind the difference between selecting local materials with potential practical value for scientific treatment and the actual attempt to adapt Western scientific methods and knowledge to the Chinese environment. On one level, it is a question of setting policies that would reflect China's own priorities and not merely the expectations of other countries. On another level, it is a question of examining and evaluating the traditions of Chinese and Western science and, if possible, integrating the two. My own study of the history of medicine in Republican China has shown that such integration was difficult. I wonder if Chinese scientists in the Republican period in fact attempted such integration on any significant scale, or if they regarded traditional science as "unscientific" and not worthy of attention.

The dependence on foreign support—both financial and intellectual—certainly contributed to the foreign character of Chinese science during

---

[2] Peter Buck, *American Science and Modern China, 1876-1939* (New York, 1980), 224-25.

[3] James Reardon-Anderson, "Science in Republican China, 1928-37," 17. Paper circulated by the author before the conference.

the Republican period. This brings to mind the Western Christian missionary experience in twentieth-century China. The Western missionary enterprise, funded and supported mainly by mission boards outside China, was attacked by many Chinese in the 1920s as a feature of Western expansionist policy. Some Chinese Christian leaders also joined in the attack, arguing that the Church in China was so dependent on foreign support that it was not taking root in China. Do we find parallels between the missionary movement and Western science in China? Were Chinese scientists as concerned as Chinese Christians about the problem of foreign dependency? How did they view and define their relationship with the American organizations and institutions that provided them with funding and support? In their view, how and when would modern science take root in China? Answers to these questions will provide us with important insights into the foreign impact on the development of sciences in the Republican period.

Finally, there is the question of Chinese scientists' reaction to the nationalistic politics of the period. During the Nanking decade (1927–37), despite the nominal unity and stability, China was in the throes of internal war between the Nationalists and the Communists and confronting an external threat in the form of Japanese aggression. How did Chinese scientists react to these problems? Did they demonstrate in the street, shouting anti-Japanese slogans, for example? Or were they insulated in their own small scientific community, believing the uninterrupted pursuit of scientific knowledge to be far more important? What role, if any, did Chinese scientists play in Republican politics? How did they define the role and function of science at a time of national crisis?

There is no doubt that the Nationalist government wanted science to serve the interests of the state, to strengthen China economically and militarily. As Reardon-Anderson has pointed out, the Kuomintang encouraged the study of applied science while discouraging the study of liberal arts subjects. It also strengthened the universities and created research institutions. But more information is needed to evaluate the role of the Nanking government in the development of science. For example, how much did the government spend on scientific research and development, and how was the money allocated? How did that amount compare with other items of expenditure? In the case of medical care, for instance, many have argued that the inadequacies of the health-care system during those years were the result of a lack of financial support from the government. Was the government willing to make large outlays in support of scientific research and development? If it was indeed willing, would such support have helped to reduce the dependence of Chinese scientists on foreign funding, and would that have led to a reduction of dependency in other areas?

The central role assigned to science in China's strengthening and construction reflects the failure of many Chinese leaders to appreciate the intrinsic value of scientific inquiry. But it is not unexpected. In a period of intense nationalistic concern for China's survival, intellectuals had used science to discredit China's traditional values and to achieve material strength. In fact, while traditional values were eventually replaced by a "scientific" system of thought, the belief that science should serve the interest of the state continues to dominate the thinking of Chinese leaders in regard to the country's modernization efforts. Indeed, in 1978, Deng Xiaoping proclaimed that "the crux of the four modernizations is the mastery of modern science and technology."[4] Science and scientists are once again called upon to rescue China from her backwardness.

---

[4] Deng Xiaoping, "Speech at Opening Ceremony of National Science Conference," *Peking Review*, No. 12 (March 24, 1978): 10.

PART II

PEOPLE'S REPUBLIC OF CHINA

PART II

PEOPLE'S REPUBLIC OF CHINA

The Role of Biomedical Research in China's Population Policy

Sheldon J. Segal

Premier Zhou Enlai tried gently to dissuade his comrades and his country from the anti-Malthusian rhetoric and policy that characterized the early postrevolutionary years of the new China. He recognized the implications of the widening gap resulting from high birth rates and declining death rates in response to improved food distribution, health care, and social organization. By 1956 Zhou was emboldened to state, in his *Proposal on the Second Five-Year Plan for National Economic Development*, that "[w]e agree to have appropriate limitations on births in order to protect our women and children; to better educate our future generations; and to benefit the nation's health and prosperity."[1] But his efforts and those of some scholars, the economist Ma Yinchu prominent among them, did not meet with full success. The formulation of an effective family planning policy was long delayed while the size of the population increased—from 542 million in 1949 to 672 million in 1959, to 803 million in 1969, to 971 million in 1979—moving inexorably

---

[1] Qian Xinzhong, "China's Population Policy: Theory and Methods," *Studies in Family Planning* 14 (1983): 295–301.

toward the 1 billion mark, which was passed shortly before 1 July 1982, when China undertook a national census.[2]

Reduction of the rate of population growth is now a sustained state policy in the People's Republic of China. The goal of the country's leaders is to keep the population at 1.2 billion and to reduce the natural growth rate to near zero by the end of this century. An official policy to promote one-child families was announced in 1979 at the Fifth National People's Congress and promulgated a year later, after considerable national discussion and debate. This extreme measure followed upon earlier attempts to encourage birth control by promoting the health and social benefits of the small family and by providing easy access to a variety of contraceptive measures, sterilization operations, and abortion.

The idea of a total fertility rate (TFR) of 1.0 is quite new in the world's demographic history. China's decision to establish this objective was not reached casually. Officials received opinions from many sources: officers of the People's Liberation Army concerned about the availability of recruits; demographers worried about the population's age structure; economists looking at labor force projections; and sociologists warning about the character flaws of the only child. Ultimately, the decision to proceed with the program was based on its being perceived as a sacrifice that the present generation must make for the sake of future generations. Policy makers are now convinced that modernization, the goal that takes highest priority in contemporary China, cannot be achieved without the tightest possible reins on population growth.[3] Zhou's prescription for China's Second Five-Year Plan has finally been filled in the Seventh Plan.

Medical and biological scientists throughout the country have been encouraged to work in the field of contraception and other aspects of human reproduction. Family planning research institutes have been established at both the state (national) and provincial levels. Institutions of the Chinese Academy of Sciences and the Chinese Academy of Medical Sciences have developed programs across a wide range of topics in

---

[2] Xu Jihui, "The World's Biggest Census," *China Population Newsletter* 1 (1983): 16–19.

[3] Fang Yi, quotation from *Agreement Contract* between the Government of the People's Republic of China and the United Nations Fund for Population Activities, Project 08 (1980).

reproductive health. In the short period since the Cultural Revolution and its stifling effect on scientific inquiry, research in reproductive biology has moved from the simple testing of traditional herbal drugs for antifertility action to solid-state protein synthesis and sophisticated procedures of molecular biology.

In 1979 the Chinese Academy of Sciences established an Institute of Developmental Biology with a mandate to modernize Chinese research in the classical field of embryology and related topics. The linkage of this undertaking to China's population policy was put forth by Fang Yi, then vice premier for science and technology. Endorsing the institute, he stated:

> Perhaps we are the only country to ask our people to have only one child per family. That gives us a special responsibility to assure that it will be a very good quality child. How can the people, especially in the rural countryside where most of our people live, have confidence that they should follow the government's advice to have one child per family unless we do everything to assure that it is a very good quality child? This is why we emphasize developmental biology and give priority to modernizing our work in this science. It is an integral part of our overall family planning programme.[4]

The institute's construction plan was given high priority by state and municipal authorities; it is now housed in a modern laboratory building in Beijing. In addition to fundamental research in developmental biology, its program includes studies on the teratogenicity of contraceptive steroids and the mutagenicity of other putative contraceptive drugs.

---

[4] Edgar Snow, *The Other Side of the River: Red China Today* (New York, 1961), 251, 413–17, 452, 586, 732.

Contraceptive Use

Visitors to China since the early 1960s have reported the widespread availability of contraceptive methods in both urban and rural areas.[5] A distinguished chemist attempted to quantify the extent of oral contraceptive use by deciphering batch number codes on labels.[6] He concluded that about 50 percent of Chinese contraceptors, in 1974, were oral contraceptive consumers, an estimate that proved to be far off the mark. Some reports characterized as generally available contraceptive preparations that were, in fact, experimental formulations with limited, local distribution and not really a significant component of the all-China pattern of contraceptive use. Thus, one heard or read reports of "vacation" pills, "weekend" pills, and "paper" pills, suggesting that China, during the isolation of the Cultural Revolution, had achieved a revolution in contraceptive techniques that had somehow bypassed the rest of the world. On closer inspection, these pills proved to be composed of well-known compounds that had, in fact, been studied elsewhere for use in the manner suggested by the Chinese applications but found to be unsatisfactory.

It was not until the census of 1 July 1982, and the subsequent, confirmatory one-in-a-thousand sample survey of September 1982, that quantitative data on contraceptive use in China became available.[7] This yielded what is probably the largest sample survey ever—1,017,544 men and women were interviewed. The 3,676 people who conducted the survey comprised a larger group than the usual sample size of surveys. The most important finding was that 69 percent of married couples of childbearing age in 1982 were using some form of contraception (including sterilization). In absolute numbers, 118 million couples, of the

---

[5] Anibal Faundes and Tapani Luukkainen, "Health and Family Planning Services in the Chinese People's Republic," *Studies in Family Planning* 3 (1972): 165–76.

[6] Carl Djerassi, "Fertility Limitation through Contraceptive Steroids in the People's Republic of China," *Studies in Family Planning* 5 (1974): 13–30.

[7] China State Family Planning Commission, "Communique of the State Family Planning Commission on a Nationwide Fertility Sampling Survey of Every Person per Thousand," *China Population Newsletter* 1 (1983): 5–7.

170 million in the reproductive age group, were practicing birth control (see fig. 1). Fifty percent of the users, or 60 million women, were depending on intrauterine contraceptive devices (IUD), making this the most widely used method of contraception not only in China but in the entire developing world. About 25 percent of the remaining women had undergone tubal ligations; and husbands had been vasectomized in about 10 percent of the defined group. Various types of pills (about 8 percent), condoms (2 percent), and other procedures (about 4 percent), including traditional Chinese herbal medicine, accounted for the remainder.

This quantitative analysis of contraceptive use establishes certain research priorities for China. It is evident, for example, that for China the purpose of contraceptive research is not primarily to find new methods that will increase the percentage of couples in the user category. The current figure of 69 percent is not significantly different from the prevailing ratios of contraceptive usage in the industrialized countries of North America or Europe.[8] In the Chinese sample survey, it was estimated that about twenty-one million couples (or 12.2 percent) who already have one child are not now using a contraceptive method. These couples, which include the minority populations who are encouraged to set their own population policies, have chosen either to have more than the prescribed single child or are prepared to accept induced abortion in the event of a pregnancy.

For China, introduction of improved contraceptive technology would reduce the number of abortions and eliminate the untoward side effects of currently used contraceptives, thereby improving the health of women, who bear the burden of the debilitating physical and mental effects of inadequate reproductive health care. That the Chinese Government recognizes this objective as it promotes contraceptive research is implied in the writings of the physician who, until his recent retirement, headed the State Family Planning Commission. Dr. Qian Xinzhong has stated, "Induced abortion may cause suffering to women and an undesirable effect on their health. We promote contraception as the main method of controlling birth because it provides better protection for the physical and mental health of women."[9]

---

[8] Dorothy L. Nortman, *Population and Family Planning Programs*, 11th ed. (New York, 1982), 74–84.

[9] Qian Xinzhong, "China's Population Policy," 299.

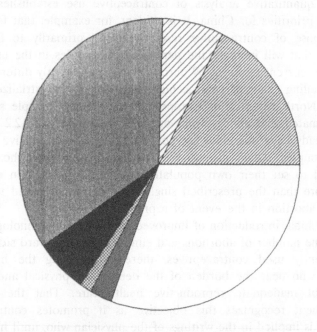

Figure 1. Contraceptive Prevalence in China, 1982.

Source: China State Family Planning Commission, "Communique of the State Family Planning Commission on a Nationwide Fertility Sampling Survey of Every Person per Thousand," *China Population Newsletter* 1 (1983): 5–7.

Chinese women had, in 1982, approximately twenty-seven million pregnancies which culminated in either live births or induced abortions (fig. 2). Of the twenty-two million live births, about 78 percent were first or second births and the remainder were of higher birth order.[10] One can assume that both the absolute number of pregnancies and the percentage of higher-birth-order pregnancies has not increased since 1982. Using existing contraceptive methods, China probably leads the world statistically in the percentage of births that are first-order births. Improved contraception, therefore, could have, at best, a minimal impact on reducing the number of unwanted higher-birth-order pregnancies.

It is through reducing the number of induced abortions—estimated to be some five million in 1982—that China can reap important benefits by the introduction of improved contraception. More than 50 percent of induced abortions in China are a consequence of contraceptive failure. The number of abortions will fall as more reliable contraceptive methods are substituted for methods that are now in use that have failure rates of three pregnancies per hundred women-years of use or higher. Although the maternal mortality rate is already admirably low in China, it will be reduced even further as the abortion rate declines.[11]

Research on Intrauterine Devices

From the analysis of current contraceptive use in China, it is evident that improving IUD performance should be a high-priority research topic for Chinese scientists. Several experimental designs have been studied,[12] but no improved model was chosen to replace the stainless steel coiled ring that is used predominantly by the sixty million Chinese women who use IUDs. The ring is a modification of the Ota ring,

---

[10] Timothy King, "Population Policy in China since 1950 and its Demographic and Economic Implications," in *Health Sector Issues in China* (Washington, D.C., 1983), 27.

[11] Katherine Ch'iu Lyle et al., "Perinatal Study in Tientsin: 1978," *International Journal of Gynecology and Obstetrics* 18 (1980): 280–89.

[12] Guangdong Family Planning Research Coordination Group, "Canton Floweran Intrauterine Device," *Chinese Journal of Obstetrics and Gynecology* 14 (1979): 283–86.

Births, 1st Order   11.2
Births, 2nd Order   6.0
Births, 3rd and Above   4.8
Abortions   8.0

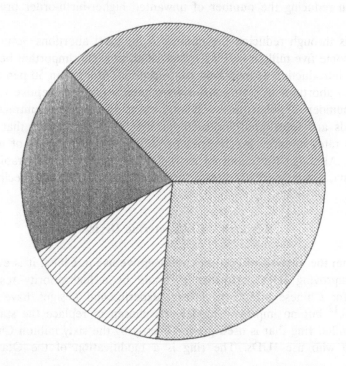

Figure 2. Pregnancy Outcome in China, 1982.
Total Pregnancies: 30 million.

Source: Timothy King, "Population Policy in China since 1950 and Its Demographic and Economic Implications," in *Health Sector Issues in China* (Washington, D.C., 1983), 27.

which received little attention when it was invented in Japan in the 1930s but became the subject of renewed interest after an encouraging 1959 publication by Ishihama in the *Yokahama Medical Journal*.[13] Although the historical record is incomplete, it was probably through the interest and intervention of Zhou Enlai that manufacture and distribution of the modified Ota ring in China began about 1963.[14]

Chinese physicians have now completed the first comparative study of the Chinese stainless steel ring (in China, called the Ma Hua ring) and two copper-bearing IUDs developed by the Population Council of New York. These T-shaped plastic devices have a record of excellent performance in several countries.[15] The Copper T-380 (manufactured in Finland), the Copper T-220 (manufactured in Mexico), and the Chinese IUD were entered into a random assignment study in five hospitals in the city of Tianjin.[16] The study design and quality of the investigation are the equal of any IUD study in the world's literature. The experience of nearly 3,000 women was analyzed by the life table procedure. Table 1 shows the number and percentage distribution of acceptors by age and parity, two key covariants that influence IUD performance. The data show that the assignment of acceptors among the three groups was successful in achieving random distribution. At the end of one year, the pregnancy rates of the Copper T-380 and Copper T-220 were 0.1 and 0.3 per 100, respectively. Both rates were significantly below the 3.0 per 100 pregnancy rate of the Chinese ring. Overall categories of performance favored the Copper T-220, which had the lowest one-year termination-of-use rate: 95 percent of Chinese women who started to

---

[13] A. Ishihama, "Clinical Studies on Intra-Uterine Rings, Especially the Present State of Contraception in Japan and the Experience in the Use of Intrauterine Rings," *Yokohama Medical Journal* 10 (1959): 89–94.

[14] In 1963 the American journalist Edgar Snow visited New York to gather information on birth control at the request of Zhou Enlai. I provided Snow with a copy of the Ishihama paper and samples of various IUDs, which he delivered to Zhou.

[15] Irving Sivin and Janet Stern, "Long-Acting, More Effective Copper T IUDs," *Studies in Family Planning* 10 (1979): 263–67.

[16] Sung Shih, Qian Li-Juan, and Liu Xuan, "Contraceptive Clinical Experience with 3 IUDs, TCu 380Ag, TCu 220C, and Ma Hua Ring, in Tianjin, People's Republic of China," *Contraception* 29 (1984): 229–39.

Table 1

Number and Percentage Distribution of IUD Acceptors by Age, Parity

Age[a]	TCu 380Ag		TCu 220C		Ma Hua	
	n	%	n	%	n	%
25–29	289	32.4	292	32.9	342	34.8
30–34	399	44.7	405	45.7	440	44.8
35+	204	22.9	190	21.4	201	20.4
Median age	32.0		31.9		31.8	
Total	892	100.0	887	100.0	983	100.0

Parity[b]	TCu 380Ag		TCu 220C		Ma Hua	
	n	%	n	%	n	%
1	572	64.1	598	67.4	688	70.0
2	283	31.7	249	28.1	265	27.0
3+	37	4.1	40	4.5	30	3.1
Median Parity	1.3		1.2		1.2	
Total	892	100.0	887	100.0	983	100.0

Source: Sung Shih, Qian Li-Juan, and Liu Xuan, "Contraceptive Clinical Experience with 3 IUD's, TCu 380Ag, TCu 220C, and Ma Hua Ring, in Tianjin, People's Republic of China," *Contraception* 29 (1984): 229–39.

[a] Chi square = 2.29, $p > .05$.
[b] Chi square = 9.17, $p > .05$.

use the Copper T-220 completed a full year of use. The comparable figure for the conventional Chinese device was 86 percent. The analysis of the results of this excellent study is shown in table 2.

It became evident to the Chinese investigators that use of either of the Copper T devices, in place of the stainless steel coiled ring, would provide major advantages to Chinese women. The incidence of cramping and irregular bleeding would be reduced; the number of accidental pregnancies with the device in place would be lower; and expulsion or removal of the device would occur less frequently. The Chinese authors estimated the annual number of IUD insertions in China to be around three million. They concluded that if either of the Copper T devices were used instead of China's conventional device, the number of accidental pregnancies would be reduced by between 78,000 and 97,000 per year. In China, an accidental pregnancy resulting from contraceptive failure usually results in voluntary induced abortion. This study demonstrated, therefore, that the technology is on hand to reduce the number of China's abortions by nearly 100,000 per year.

This important study was done in the context of the United Nations Fund for Population Activities (UNFPA) Country Program in agreement with the government of the People's Republic of China. The UNFPA program, which is implemented by the State Family Planning Commission, includes a project for the transfer of contraceptive technology for local manufacture in China. Under this project, the manufacturing capability for the Copper T-200 device has already been introduced in Tianjin. Three hundred thousand of the new devices had been manufactured by March 1984, and the facility was being scaled up to produce three million devices a year. It has been possible, therefore, for China to translate into action the programmatic implications of this valuable comparative study of IUDs.

Research on Steroidal Contraceptives

China has sophisticated capabilities in chemical synthesis, and the chemical industry is particularly competent in the steroid field. In fact, China has entered the international market as an exporter of bulk steroids, using diosgenin obtained from indigenous yams as the starting

Table 2

First Segment Net Cumulative Rates per 100 at Completion of Twelve Months

Reason for Termination	Rates (S.E.)					
	TCu 380Ag		TCu 220C		Ma Hua	
Pregnancy[a]	0.1	(0.1)	0.3	(0.2)	3.0	(0.6)
Expulsion[a]	4.3	(0.7)	1.6	(0.4)	6.3	(0.8)
Medical reasons						
Bleeding or pain[b]	6.8	(0.9)	3.3	(0.6)	4.3	(0.7)
Other[c]	0.3	(0.2)	0.0	(--)	0.3	(0.2)
Personal reasons						
Planning pregnancy[c]	0.1	(0.1)	0.3	(0.2)	0.0	(--)
Total terminations[a]	11.6	(1.1)	5.5	(0.8)	13.9	(1.1)
Continuation/ completed year[a]	88.4	(1.1)	94.5	(0.8)	86.1	(1.1)
Number of woman years	810		799		874	
Total acceptors	892		887		983	

Source: Sung et al., "Contraceptive Clinical Experience in Tianjin."

[a] $p < .001$.
[b] $p < .05$.
[c] $p < .05$.

material.[17] By 1982 China had gained a significant share of the bulk steroid market, but disastrous floods in the yam-producing regions of China in 1982–83 made it impossible to meet delivery schedules. Consequently, many buyers returned to the more costly suppliers who use total synthesis techniques.

With the availability of competence in steroid synthesis, China has had access to samples of virtually all of the contraceptive steriods either tested experimentally or used commercially in Western countries. It is not clear whether Chinese synthesis of these compounds arose *de novo* or followed publication of the structures in the international scientific or patent literature.[18] China's record with regard to steroidal contraception is paradoxical. On the one hand, the country's scientists have made ingenious use of the characteristics of specific compounds to fit China's particular social situations. On the other hand, steroidal contraception has had little overall impact in China. Users of all forms of steroidal contraceptives comprise fewer than 10 percent of couples using contraception or sterilization (fig. 1).

China's conventional contraceptive "pills"—taken daily for three weeks out of the month—are comprised of the same progestins and estrogens used in similar products elsewhere. China preceded the rest of the world in using lower dosages of contraceptive steroids. As early as 1969, combination pills of 625 micrograms of norethisterone and 35 micrograms of ethinyl estradiol, or 300 micrograms of norgestrel and 30 micrograms of ethinyl estradiol, were being distributed. Comparable low-dose preparations were not introduced in the United States until five years later.[19] Thus, while Chinese women were being spared the overdose, perhaps tons of unnecessary steroids were passing through the

---

[17] Carl Djerassi, "Fertility Limitation through Contraceptive Steroids," 20.

[18] In 1980, I was given by Chinese colleagues a flow chart showing the steps used to synthesize a steroid identical in structure to a progestational agent produced by Roussel-UCLAF of Paris. With permission, I shared this with Roussel chemists at their laboratories at Roumainville, outside of Paris. When I asked the head chemist at Roussel if they synthesize it the same way, he replied, "No, but this is the way we *published* that it could be synthesized."

[19] Djerassi, "Fertility Limitation through Contraceptive Steroids," 20.

hepatic portals of women in the rest of the world. Chinese gynecologists, in fact, tried to reduce dosages of combination pills even further, but when the dose of norethindrone, for example, fell below 500 micrograms, bleeding control was inadequate to provide an acceptable product. The record of how these dosage calibrations were worked out is not evident in the Chinese medical literature. The deputy director of the National Research Institute for Family Planning in Beijing reported this transition from high- to low-dose combinations in a review paper published in 1984,[20] but no references to the primary data sources were included.

China manufactures an injectable contraceptive steroid, designed to be administered once a month. It is a combination product identical to one that was once marketed by Schering AG of Berlin but discontinued in the early 1960s because of poor performance. There is no evidence that the monthly injectable, which contains 17-alpha-hydroxy progesterone caproate and estradiol valerate, is more popular with Chinese women than it was with women of Central and South America, where it was marketed twenty years ago. Other injectable contraceptives are being tested in China, thanks to the cooperation of the World Health Organization.[21]

Chinese scientists have attempted to adapt the properties of steroidal contraception to the needs of those Chinese couples who experience long periods of separation and only short periods of cohabitation. The generic term "vacation pill" is employed to describe several preparations with different modes of action. It is, indeed, desirable to avoid continuous use of steroids when coital exposure is confined to a single ovulatory cycle during the year. It is difficult to understand, however, why it would be preferable to use steroid compounds at high doses that can produce a variety of undesirable side effects when the option exists to use the conventional pill for one or two months. The national survey data (fig. 1) suggest that these nonconventional usages of steroid com-

---

[20] Xiao Bilian, "Some Methods of Fertility Regulation in China: Experience and Problems," in *Research on the Regulation of Human Fertility: Needs of Developing Countries*, eds. Anne and Egon Diczfalusy (Copenhagen, 1984), 120–40.

[21] Peter E. Hall and Ian S. Fraser, "Monthly Injectable Contraceptives," in *Long-Acting Steroid Contraception*, ed. D. R. Mishell, Jr. (New York, 1983), 65–88.

pounds have not had much impact on the pattern of contraceptive use in China.

China's most significant research initiative in steroidal contraception is a thirty-month program which began in June 1984 to test a long-acting contraceptive subdermal implant for possible inclusion in China's nationwide distribution system. The NORPLANT® contraceptive is the product of an international research effort coordinated by the Population Council and has been approved for manufacture and commercial distribution in Finland. Testing continues in the United States and elsewhere, including extensive field studies in Indonesia, Thailand, and Egypt. The State Family Planning Commission has embarked on China's own program, which will provide for testing of the new contraceptive in both urban and rural areas; training of personnel in the use of implant contraception; development of informational material for potential users; and, most importantly, the transfer of technology to insure manufacture of NORPLANT® implants in China.

NORPLANT® implant use provides women with almost complete protection against pregnancy. Studies reveal that the pregnancy rate is less than 0.5 per hundred women-years of use.[22] Once inserted, the implants extend five years of protection but can be removed at any time so that fertility is restored. Otherwise, a new set of implants can be inserted so that two visits to a clinic can provide a full decade of protection. The level of protection and the long-acting feature of the method give it the characteristics of a reversible sterilization procedure or a contraceptive "vaccine."

Implant contraception may prove to have special merit in China, where couples have already demonstrated their preference for long-acting methods. As noted previously, 85 percent of contraceptive users in China select IUDs, tubal ligation, or vasectomy. NORPLANT® implants provide an option for reversibility not available otherwise, particularly for those couples who might otherwise choose sterilization operations. The method has already been found to have appeal and high-performance characteristics in other developing countries where testing is under way. How it will be received by Chinese women will be known in a relatively short time, thanks to the ability of the Chinese system to coordinate research activities with policy decisions.

---

[22] Sheldon J. Segal, "The Development of NORPLANT® Implants," *Studies in Family Planning* 14 (1983): 159–63.

Research on Gossypol as a Contraceptive Pill for Men

Scientists have been seeking safe pharmacological means to interfere with spermatogenesis. Candidate compounds produced thus far have either been too toxic or too nonspecific, suppressing the testis's hormone-producing function as well as its gamete-producing function. Thus, considerable interest was elicited by a report from China in 1978 of extensive clinical trials using a compound that could accomplish that elusive goal of preventing the production of sperm without either stopping the production of testosterone or suppressing libido.[23] The Chinese report was greeted with some skepticism. An official of the National Institutes of Health in Bethesda was quoted as saying, "Historically, there have been other occasions where the Chinese have made claims of clinical efficacy which could not be substantiated."[24] But this skepticism did not prevent the Chinese experiment from stimulating a considerable amount of scientific work with gossypol, the compound studied in China by the National Coordinating Group on Male Antifertility Agents.

Gossypol is a phenolic compound which can be extracted as a yellowish powder from various parts of the cotton plant. The extraction process is simple and inexpensive, and the gossypol produced can be taken to a high degree of purity by routine procedures. Its antifertility action was discovered serendipitously in China in the late 1960s, when physicians were attempting to discover the cause of an epidemic of infertility in Hebei, Jiangsu, and other cotton-producing provinces. They found that the affliction was associated with the ingestion of crude cottonseed oil. The clue that turned attention to uncooked cottonseed oil was found in a long-overlooked paper published in a Shanghai medical journal in 1957.[25] That paper described the clinical observations of

---

[23] National Coordinating Group on Male Antifertility Agents, "Gossypol—A New Antifertility Agent for Males," *Chinese Medical Journal* 4 (1978): 417–28.

[24] "Visitors Say Chinese Report They Have A Safe and Effective Birth Control Pill For Men," *The New York Times*, 6 January 1979.

[25] Liu Ban-shan, "Suggestions of Feeding Crude Cottonseed Oil for Contraception," *Shanghai Medical Journal* (1957).

a practitioner of traditional herbal medicine. He wrote that many of his patients, both male and female, were unable to conceive if they used crude cottonseed oil in their daily diet.

Several research groups in China, particularly in Beijing, Shanghai, and Nanjing, undertook studies in laboratory animals, and soon the identification of gossypol as the sterility-causing agent was made and confirmed. By 1973, with the discontinuation of the use of crude cottonseed oil, fertility was returning in the afflicted rural villages, and this added evidence to animal studies which had by then revealed that the effects of gossypol could be reversible. The laboratory work turned to other facets of the problem. To make the compound more stable and less susceptible to deterioration caused by exposure to light, acetic acid was attached to the gossypol molecule. With excellent consistency, each new experiment with laboratory rats confirmed the earlier findings that gossypol or gossypol monoacetate, like crude cottonseed oil, causes male infertility. In a short time confirmatory studies were done around China using rabbits, pigs, dogs, and monkeys.[26]

The decision to initiate clinical trials of gossypol as a male contraceptive was facilitated by the fact that alcoholic extracts of cotton root containing gossypol were being used in Chinese traditional herbal medicine for treating bronchitis. A few small-scale studies were carried out in 1971 and 1972, involving first patients with bronchitis and then volunteers willing to test the drug as a contraceptive. This phase of the work established that daily doses of 20–35 mg gossypol cause azoospermia after five to seven weeks of treatment. On the basis of these preliminary studies, a multicenter program was established by the National Coordinating Group on Male Antifertility Agents and gossypol underwent extensive clinical trials throughout China. The results reported in 1978 aggregated the data from all of the fourteen participating centers. Different dosage regimens had been used, each center had prepared its own drug, and patient selection and follow-up were not standardized. The Chinese investigators were fully aware of these limitations, but the results were so impressive that the National Coordinating Group decided to publish them and provided an English version that would be accessible to colleagues outside of China. In all, 8,806 volunteers had been studied over a span of seven years. The data were

---

[26] Xue Shepu, "Studies on the Antifertility Effect of Gossypol, A New Contraceptive for Males," *Scientia Sinica* 26 (1983): 614–33.

reported in terms of percentage of men with sperm counts of four million per cubic centimeter or lower. Side effects were also reported in terms of percentages.

From the data presented it was concluded that the oral administration of 20 mg of gossypol per day for seventy-five consecutive days is highly effective in suppressing spermatogenesis and that a maintenance dose of 50 mg per week will maintain the suppression. With respect to safety issues, blood chemistry and other clinical pharmacological tests had been carried out on many of the men. A reduction in serum potassium to below-normal levels was observed in sixty-six cases. These cases were grouped in some centers, while other centers reported no changes in potassium levels. Incidence data relating to subjective symptoms such as fatigue, change in libido, loss of appetite, and headaches were also obtained. No untoward effects of particular concern were revealed. In some cases plasma testosterone levels had been determined and found to be normal.

On the basis of these findings, in early 1980 the National Coordinating Group recommended to the Ministry of Public Health that gossypol as a contraceptive pill for men be introduced for general use. Some members of the group dissented, calling for further research to resolve the issues of hypokalemia and reversibility. The latter opinion prevailed and, consequently, the Capital Medical College of China (CMCC) in 1980 initiated a new study of gossypol, aimed at resolving some of the issues in question. The effect of the drug on serum potassium levels is of particular concern in the CMCC study.

Because gossypol users are advised to use other forms of contraception during the seventy-five day loading phase, it was ethically possible to introduce a placebo group and design the loading phase of the study as a double-blind, randomized, controlled clinical trial.[27] Investigators in the urology department of Capital Hospital recruited 152 participants in two groups: one group of 75 volunteers received 20 mg per day of gossypol and 77 volunteers received placebo pills. The success of randomization is shown by the physiological and demographic characteristics of the two groups detailed in table 3. At the end of the loading phase, there was no statistical difference in serum potassium values between

---

[27] Liu Guozhen, Katherine C. Lyle, and Cao Jian, "Clinical Trial of Gossypol as a Male Contraceptive Drug. Part I. Efficacy Study," *Fertility and Sterility* 48 (1987): 459–61.

Table 3

CMCC Gossypol Study, 1982–1984

A. Physiological Match

Characteristic	Gossypol (n = 75)		Placebo (n = 77)	
	Mean	S.D.	Mean	S.D.
Age	31.7	4.2	31.0	4.2
Weight	64.6	8.3	62.6	8.2
Coital frequency (per week)	1.5	0.5	1.6	0.5
Blood pressure				
Systolic (mmHg)	116.1	11.2	114.4	11.8
Diastolic (mmHg)	82.1	8.4	80.5	9.1
Hemoglobin (gm/100ml)	15.7	1.2	15.5	1.2
Serum potassium (mEq/L)	4.25	0.43	4.32	0.48
Sperm count (10^6/ml)	80.0	47.0	88.0	50.0

B. Demographic Match

Characteristic	Gossypol (n = 75)	Placebo (n = 77)
Age (mean)	31.7	31.0
Education (years)		
1–6	11	9
7–9	57	57
10–12	3	9
13+	4	2
Occupation		
Worker	70	71
Office	5	5
Medical/professional	0	1

Source: Liu Guozhen and Katherine C. Lyle, "Trial of Gossypol as a Male Contraceptive," in *Gossypol*, ed. S. J. Segal (New York, 1985), 11.

the gossypol and placebo groups (table 4). This encouraging finding in clinical chemistry was reinforced by the clinical observation that none of the participants complained of symptoms associated with hypokalemia. However, through the ensuing twelve months of the maintenance phase, a statistically significant trend toward reduced serum potassium levels continued (table 5).[28]

The CMCC study confirmed the initial Chinese finding that gossypol suppresses sperm production without inhibiting libido. With this study, the investigators provided the first life-table analysis of an oral contraceptive used by men (table 6). The study presented a statistically valid analysis of the effect of gossypol, at the dosages studied, on serum potassium levels. It has demonstrated that the effect was minimal and had no apparent clinical significance in the cases studied. From these results, a new study has been designed and initiated to study the effects of a lower dose and to evaluate the merits of ingesting potassium salt while using gossypol.

Conclusion

In this paper I have described some examples of biomedical research which have important implications for China's birth control program and population policy. The comparative study of IUDs conducted in Tianjin is the first step toward replacing a technology that has been used in China for more than twenty years. The new IUD technology that the Chinese will introduce can reduce by 90 percent the accidental pregnancies associated with intrauterine contraception. At the estimated rate of IUD acceptance in China, this technological advance could reduce by one hundred thousand per year the number of induced abortions requested because of contraceptive failure.

The Capital Medical College of China's study of gossypol is an example of the application of sophisticated experimental design to clinical research. The results have enabled the Ministry of Public Health to break an impasse with respect to the role of gossypol as a possible pill for men in the Chinese program. In time, gossypol could provide

---

[28] Liu and Lyle, "Clinical Trial of Gossypol as a Male Contraceptive Drug. Part II. Hypokalemia Study," *Fertility and Sterility* 48 (1987): 462–65.

Table 4

CMCC Gossypol Study, 1982-1984

Serum Potassium at Admission and at the End of the Double-Blind Phase (20mg/day for 75 days)

	Gossypol (n = 64)	Placebo (n = 74)	T (Independent Sample)
At admission	4.25 ± 0.43	4.32 ± 0.48	.9($p > .05$)
At end of double-blind phase	4.16 ± 0.42	4.18 ± 0.41	-.3($p > .05$)
T (paired-sample)	1.6 ($p > .05$)	2.0 ($p < .05$)	

Source: Liu and Lyle, "Trial of Gossypol as a Male Contraceptive."

Table 5

CMCC Gossypol Study, 1982-1984

Serum Potassium Levels ($F = 30.0305$, $p < .05$)

Time	Number	Mean Level (mEq/L)	Standard Error	95% Confidence Limit
At admission	152	4.29	0.03	4.23-4.35
At end of loading	124	4.06	0.04	3.99-4.14
At 3 months	118	3.93	0.04	3.85-4.01
At 6 months	108	3.82	0.04	3.74-3.90
At 9 months	95	3.86	0.04	3.78-3.94
At 12 months	86	3.75	0.04	3.67-3.83

Source: Liu, Lyle, and Gao Jian, "Clinical Trial of Gossypol as a Male Contraceptive Drug. Part I. Efficacy Study," *Fertility and Sterility* 48 (1987): 459-61.

Table 6

CMCC Gossypol Study, 1982–1984

Net Cumulative Termination and Continuation Rates after Six and Twelve Months of Use

Reason for Termination	Six Months	Twelve Months
Pregnancy	0.0 ± 0.0	0.0 ± 0.0
Medical reasons		
Fatigue	1.3 ± 0.13	3.3 ± 0.06
Decreased appetite	4.0 ± 0.07	4.6 ± 0.09
Decreased libido	0.7 ± 0.004	1.3 ± 0.014
Other	2.6 ± 0.04	5.3 ± 0.11
Method-related	10.5 ± 0.29	15.8 ± 0.7
Personal reasons	5.3 ± 0.11	2.0 ± 0.02
Technical	1.3 ± 0.01	2.0 ± 0.02
Total Termination	25.7	38.3
Continuation	74.3 ± 3.52	61.2 ± 4.07
Man months	771.5	1393.5
Lost to follow-up	0	0
Total Enrolled	152	

Source: Liu et al., "Clinical Trial of Gossypol (I)."

a nonsurgical, reversible alternative to vasectomy. More information is needed, and the new study is already under way to provide answers.

This review is not intended to be comprehensive. There are interesting Chinese initiatives in the fields of reversible sterilization procedures and abortifacients which have not been covered. These are exceedingly important research topics in a country where low parity is the rule.

In the wake of policies encouraging late marriage, delayed birth, and the one-child family, it seems inevitable that the number of abortions in China will rise unless more effective contraception is found and made available to the people. This presents a major challenge to both biomedical scientists and policy makers in China, who must move quickly to implement improvements and innovations in contraception.

"Farewell to the Plague Spirit": Chairman Mao's Crusade against Schistosomiasis

Kenneth S. Warren

In July 1958 Mao Zedong wrote a poem to commemorate the elimination of schistosomiasis in Yujiang County (located south of the Yangzi in Jiangxi Province). The poem begins with the prose statement, "Reading in the *People's Daily* of June 30, 1958, about the stamping out of the blood-fluke epidemic in Yukiang [Yujiang], my mind became so turbulent I could not sleep. A soft breeze blew warmly and the rising sun sparkled on the window. Looking far into the southern sky, I began to write."

> Green water, blue mountain—beautiful in vain.
> Even that ancient doctor, Hua To,
> could not have stopped these tiny worms.
> Hundreds of villages over-grown with weeds.
> Sick people shitting.
> Thousands of desolate houses—ghosts singing.
> Merely sitting here, each day
> we travel eighty thousand miles on the turning earth.
> Exploring the sky—looking at a thousand Milky Ways.
> The cowherd asked what to do about the Plague Spirit:

the same sorrow and joy still
float on the flowing waves.

Spring wind—willow leaves in thousands.
Six hundred million in China—
all Emperor Yao and Shun.
Red rain of blossoms whirling in waves.

Blue mountains by hard work turned into bridges.
Heaven-touching Five Peaks, pickaxes falling.
Three rivers trembling, iron arms shaking.
May we ask Mr. Plague—where do you want to go?
Paper boats on fire, candles lit, the sky burning.

This translation was done by Hua-ling Nieh Engle and Paul Engle, who provided the following commentary: "Yukiang had special meaning for Mao because he had been there with the Red Army during hard fighting against the Nationalists in the late 20's and early 30's in that area." Engle further stated that "[n]o poem of Mao's has a wider variety of influences and references to the Chinese experience, past and present, than this one. Schistosomiasis thus becomes a symbol of Mao's effort to stamp out all that was bad in the past of China."[1]

Although Mao led many crusades, the one against schistosomiasis clearly was of particular importance to him. This was emphasized during an interview in 1982 with Su Delong, vice rector of Shanghai First Medical College.[2] Professor Su had worked on schistosomiasis in a highly endemic village in Jiangxi as early as 1942. In 1949, when the People's Liberation Army freed Shanghai, soldiers from other areas who had previously been unexposed to schistosomiasis developed severe diarrhea, fever, and weakness. A group was formed to deal with the problem and Dr. Su was appointed vice secretary. Later he studied schistosomiasis around Shanghai and found a very high rate of infection. In 1955 Chairman Mao showed a particular interest in the problem, and in 1956 an official commission was organized to deal with it. Asked why

---

[1] H. N. Engle and P. Engle, *Poems of Mao Tse-tung* (New York, 1973), 101–6.

[2] C. H. Warren, "Recorded Interview with Professor Su Delong," March 1983 (personal communication).

Mao was interested, Dr. Su replied, "Because he deals with so many of the people and soldiers suffering from this disease." On 7 July 1957 Dr. Su was summoned by the chairman, who asked him to describe the situation and review what could be done about it. After hearing Su's report, "Mao's response was, 'Well, can we eradicate this disease within seven years?' I [Su] said I didn't know, perhaps the time is too short. He changed his idea. He said, 'How about twelve years? It is slightly better. It is better.' He issued an order saying this disease must be wiped out in twelve years. At that time, his order was heard, high authority, you see. Everybody has to obey his orders."

Schistosomiasis is among the most ancient of all verified diseases of mankind. (See table 1 for a chronology of schistosomiasis and its control in China.) In addition to the demonstration of schistosome eggs in Egyptian mummies, *Schistosoma japonicum* eggs have recently been found in Chinese "corpses" disinterred in Hunan and Hubei provinces and dating back two thousand years. Old Chinese writings presumably described schistosomiasis long before Fuji's classical description of Katayama fever (a severe, sometimes fatal influenza-like illness) in Japan in 1847. For example, in 400 B.C. a "water poison attacking man . . . like a poisonous insect but invisible" was described. In the seventh century it was noted that

> in hill regions south and east of three *wu* (corresponding to Suzhou in Jiangsu Province and Wuxing and Shaoxing in Zhejiang Province), there is water poison disease in streams contracted easily in spring and autumn . . . in water, there are minute and invisible sand lice that penetrate human skin when man is bathing in streams or in water from the streams . . . rash appears at the beginning, in the size of millet and is prickling upon touch . . . onset with chill, headache and orbit pain.[3]

It was not until 1905, however, that O. T. Logan reported "A Case of Dysentery in Hunan Province Caused by the Trematode *Schistosoma*

---

[3] Mao Shoupai and Shao Baoruo, "Schistosomiasis Control in the People's Republic of China," *American Journal of Tropical Medicine and Hygiene* 31 (1982): 92–96.

Table 1

Chronology of Schistosomiasis and Its Control in China

100 B.C.	Schistosome eggs found in corpses.
A.D. 610	Putative description of schistosomiasis.
1905	*Schistosoma japonicum* infection first described in China.
1924	Faust and Meleney publish *Studies on schistosomiasis japonica.*
1950	Institute of Parasitic Diseases founded in Shanghai.
1951–56	National survey finds 10,470,000 Chinese infected.
1955	Japanese delegation produces *Recommendations for Control* (published in 1957).
1955	Nine-man "Leading Group" formed by Central Committee of the Party.
1958	Mao Zedong writes "Farewell to the Plague Spirit."
1958	All-China Conference on Parasitic Diseases calls for eradication within one year.
1958–60	Great Leap Forward; commune system developed.
1967–70	Cultural Revolution; control inhibited.
1979	*New York Times* reports a new eradication attempt to succeed by 1985.
1982	Institute of Parasitic Diseases publishes *Schistosomiasis Control in the People's Republic of China,* reporting one-third of area still endemic with approximately 2.5 million cases.
1983	First quantitative studies of human schistosomiasis appear in the *New England Journal of Medicine.*

japonicum" in the *China Medical Missionary Journal*.[4] This was only one year after the discovery in Japan of *Schistosoma japonicum* in a cat and soon thereafter in man.[5] The life cycle of the parasite was elucidated in Japan in 1913.[6] Until 1924 virtually all of the studies in China were essentially case reports, with a few investigations of the distribution of infection.[7] That year, Faust and Meleney's exhaustive 339-page monograph, *Studies on Schistosomiasis japonica*, appeared. Their investigations were performed in China and included research on the morphology and biology of the developmental stages of the worms; the pathological effects of the worms on both their human and molluscan hosts; the distribution of schistosomiasis in China, including the environmental conditions governing such distribution; and the treatment and control of the infection. With respect to distribution, a questionnaire sent to Western doctors throughout the country revealed that nine of thirteen suspected provinces had definite areas of infection; most importantly, the Yangzi flood area. Faust and Meleney studied the infection in experimental rabbits and dogs and also intensively examined seventeen hospitalized patients with advanced hepatosplenic disease. They described emaciation, enlarged liver and spleen, ascites (fluid in the abdomen), dilatation of the veins of the abdomen, anemia, and hematemesis (vomiting of blood). The treatment of choice at the time was toxic antimony drugs.[8]

From 1924 until 1958 the literature on schistosomiasis was relatively small (fewer than two hundred papers) and was concerned largely with case reports and distribution studies. During the Great Leap Forward

---

[4] O. T. Logan, "A Case of Dysentery in Hunan Province Caused by the Trematode *Schistosomum japonicum*," *China Medical Missionary Journal* 19 (1905): 243–45.

[5] F. Katsurada, "*Schistosomum japonicum*, A New Parasite of Man, by Which an Endemic Disease in Various Areas of Japan is Caused," *Annotationes Zoologicae Japanenses* 5 (1904): 146–60.

[6] K. Miyairi and M. Suzuki, "On the Development of *Schistosoma japonicum*," *Tokyo Medical Journal*, no. 1836 (1913): 1–5.

[7] K. S. Warren and V. A. Newill, *Schistosomiasis: A Bibliography of the World's Literature from 1852–1962* (Cleveland, 1967).

[8] E. C. Faust and H. E. Meleney, "Studies on Schistosomiasis japonica," *American Journal of Hygiene*, Monograph series 3 (1924): 1–39.

period (1958-60) a period of intense study of schistosomiasis was undertaken by teams of investigators throughout China. The marked increase in publications on the disease during that period is shown in figure 1.[9]

Concern about schistosomiasis and other parasitic diseases led to the establishment in 1950 by the Chinese Academy of Medical Sciences of the Institute of Parasitic Diseases in Shanghai. From 1951 to 1956, mass surveys were conducted using both stool examinations and skin testing and 10,407,000 infections were reported. Severe schistosomiasis, in which entire areas were devastated, was also reported, primarily in the Shanghai region. Villages were describing mortality rates of 50 to 75 percent, with more than 90 percent infection rates. The famous "Village of the Widows" now contains a museum and a shrine to the victims of this disease. Graphic descriptions were provided:

> Tzeshih village of Kiangning county, Hupei province, where the disease levied such a heavy toll on the inhabitants that 12,000 *mu* (*mu* = one-sixth of an acre) of fertile land which once made the village famous for rice production, were overgrown with weeds. Scourged by famine, the people survived on seaweed in spring, wild herbs in summer, husks in autumn, and handouts in winter. Many ended the nightmare of their existence with suicide.[10]

In 1955, the Chinese Government, which had noted the great degree of control of schistosomiasis already achieved in Japan, asked the Japanese Society of Parasitology to advise them on control methods. A Japanese medical delegation headed by Professor Y. Komiya spent almost two months in China; its recommendations were published in 1957. These put a major emphasis on prevention, the principal aim of which was elimination of the snail vector: "Environmental changes consist 1) in displacement of soil to bury the snails, 2) covering the river and lake banks with stones, 3) constructing river banks out of concrete material

---

[9] W. Goffman and K. S. Warren, *Scientific Information Systems and the Principle of Selectivity* (New York, 1980), 50.

[10] T. H. Cheng, "Schistosomiasis in Mainland China, A Review of Research and Control Programs since 1949," *American Journal of Tropical Medicine and Hygiene* 20 (1971): 26-53.

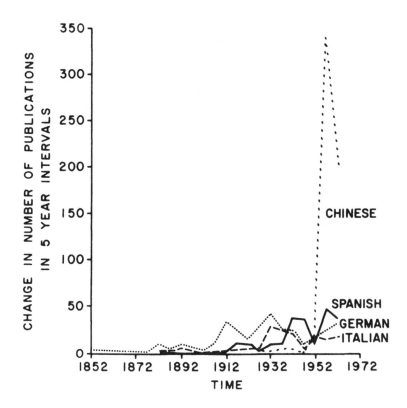

Figure 1. Articles on Schistosomiasis, 1852–1972.

Source: W. Goffman and K. S. Warren, *Scientific Information Systems and the Principle of Selectivity* (New York, 1980), 50. Reprinted by permission of Praeger Publishers.

and 4) land reclamation."[11] This approach became the major control strategy of China, although it has also been claimed that molluscicides were used extensively.[12] Late in 1955, the antischistosomiasis campaign was removed from the control of the Ministry of Public Health and put in the hands of a nine-man "Leading Group" set up by the Central Committee of the Party. A major approach introduced by that group was mass chemotherapy: in 1956 and 1957, some nine hundred thousand persons were treated with sodium antimony tartrate, essentially the same treatment described by Faust and Meleney in 1924. A three-day intensive course introduced in 1958 was reportedly used on as many as five million infected persons between 1958 and 1962.[13]

In 1982 a review of major importance, entitled "Schistosomiasis Control in the People's Republic of China," by Mao Shoupai and Shao Baoruo of the Institute of Parasitic Diseases, was published in the Western literature.[14] The authors provided a review of the state of disease distribution before control efforts, the control measures used, and present status of schistosomiasis. The distribution of schistosomiasis in three main ecological areas of the country was described. The first area, defined as the densely populated plain region with enriched networks of water sources in the delta of the Yangzi, including Shanghai and part of Jiangsu and Zhejiang provinces, was found to contain 33.6 percent of the infected people, even though it represented less than 10 percent of the snail-ridden area of the country. The second area, the mountainous and hill regions, also comprising about 10 percent of the snail-infested area, turned up some 22.7 percent of the total victims. And the third, described as the marsh and lake regions of the country, including all the areas on the banks of the Yangzi and surrounding lakes of varying size in Hunan, Hubei, Jiangxi, Anhui, and Jiangsu provinces, comprised by far the greatest area of possible infection (82 percent). More sparsely populated than the other areas, this region was found to contain approximately 44 percent of the infected individuals.[15]

---

[11] Y. Komiya, "Comments and Recommendations Concerning the Control of Schistosomiasis in China," *Japanese Journal of Medical Sciences and Biology* 10 (1957): 461–71.
[12] Warren, "Interview with Professor Su Delong."
[13] Cheng, "Schistosomiasis in Mainland China."
[14] Mao Shoupai and Shao Baoruo, "Schistosomiasis Control."
[15] Ibid.

In retrospect, snail control was one of the most important programs undertaken by the People's Republic. It was closely integrated with agricultural production, and molluscicides were used only as a supplementary measure. Vast numbers of people were mobilized to bury snails, largely by filling in irrigation canals completely and building new canals beside them. In the marshy regions, land reclamation projects were initiated. Treatment was also a major aspect of the control program, the antimony drugs serving as the mainstay. The treatment campaigns occurred at every level in the country, from the barefoot doctors up to the almost two hundred hospitals created specifically to treat schistosomiasis victims. Because *Schistosoma japonicum* also infects a wide variety of animal reservoirs, hundreds of thousands of cattle, the most important reservoir in China, were also treated in the control campaigns.[16]

Through the cacophony of the eradication campaigns and their claims of success within various time frames (one, seven, and twelve years), it has been difficult for outsiders to assess the success of this program. From the very beginning experts on schistosomiasis had grave doubts that such a program could succeed; Schistosomiasis japonica is theoretically the most difficult of all the forms of schistosomiasis to control. First, the snails are amphibious rather than purely aquatic. Second, there are a large number of different reservoir hosts, both domestic and wild; and third, it is the most difficult form of schistosomiasis to treat, requiring large amounts of highly toxic drugs. A major assessment of the programs was provided in 1971 by T. H. Cheng, who noted the frequent claims of spectacular successes in the Chinese literature. He described a series of highly successful localized campaigns but concluded, "In my opinion, China's struggle against schistosomiasis is by no means won and her vaunted seven-year plan for nationwide schistosomiasis eradication has fallen behind schedule."[17]

In 1975 a delegation of American experts visited the People's Republic of China to assess the schistosomiasis control campaign. Their conclusion was, "Unfortunately, we were unable to evaluate with con-

---

[16] Ibid.
[17] Cheng, "Schistosomiasis in Mainland China."

fidence the degree to which transmission has been reduced."[18] In 1981 I visited the Institute for Parasitic Diseases in Shanghai and was astonished to see a large-scale map showing the distribution of schistosomiasis in China before and after the eradication efforts. This map was reproduced in the 1982 paper by Mao and Shao.[19] A more recent map is presented here as figure 2. It appears that there has been a reduction of approximately two thirds of the infected area, with about one quarter of the original number of people still remaining infected.

In 1981 plans were made for an international group of scientists to perform a study of clinical schistosomiasis in an area of China in which major control measures had not been instituted. This was in the marshy regions around the Yangzi River, in which snails are very widely distributed and the population is relatively sparse. A unique aspect of this investigation was the introduction into China, for the first time, of quantitative egg-counting methods to establish the intensity as well as the prevalence of infection and to study the relationship between worm burden and the development of disease.[20] Schistosomiasis japonica has been reputed to be the most severe of all of the human schistosome infections because of the high recovery of worms following exposure to infective larval forms and the exceedingly high egg output by the worms. Studies in experimental animals have revealed, however, that in spite of the enormous numbers of pathogenic eggs emitted by *S. japonicum*, survival is high and the disease undergoes amelioration with time.[21] Quantitative investigations of human schistosomiasis japonica performed in an uncontrolled area of Leyte in the Philippines revealed moderate prevalence, relatively low intensity, and very little evidence of chronic

---

[18] "Report of the American Schistosomiasis Delegation to the People's Republic of China," *American Journal of Tropical Medicine and Hygiene* 26 (1977): 427–57.

[19] Mao Shoupai and Shao Baoruo, "Schistosomiasis Control."

[20] K. S. Warren, D. L. Su, Z. Y. Xu, H. C. Yuan, P. A. Peters, J. A. Cook, K. E. Mott, and H. B. Houser, "Morbidity in Schistosomiasis japonica in Relation to Intensity of Infection: A Study of Two Rural Brigades," *New England Journal of Medicine* 309 (1983): 1533–39.

[21] K. S. Warren and E. G. Berry, "Induction of Hepatosplenic Disease by Single Pairs of Philippine, Formosan, Japanese and Chinese Strains of *Schistosoma japonica*," *Journal of Infectious Diseases* 126 (1972): 482–91.

disease.[22] These surprising studies provided an impetus to examine the disease in China in a locale where it had been reported to be of particular severity.

The investigations in China were performed by American and Chinese biomedical scientists and epidemiologists in a rural commune in Guichi County, Anhui Province. Two different brigades of the commune were studied separately, brigade A containing 778 persons and brigade B, 1,532 persons. The prevalence in the former was 26.3 percent and in the latter, 14.4 percent. Intensity was relatively low and similar to that seen in the Philippines. Since disease tends to occur only in those with severe infections, it is interesting to observe that this was seen in only 2 percent of those examined (fig. 3). Clinical symptoms were compared among uninfected brigade members and those at three levels of infection as determined by egg output. Some increased weakness was seen only at the heaviest levels of infection; abdominal pain was not an important symptom. Liver enlargement was somewhat more frequent in moderate and severe infections, but spleen enlargement was rare and unrelated to intensity of infection. Again, the results were similar to those seen in the Philippines. Not a single person with severe hepatosplenic schistosomiasis was found. While this may have been related to some degree of land reclamation and use of molluscicides, the most striking finding was that 50 percent of all of the people in the brigades had been treated one or more times with antimony potassium tartrate, with 12 percent receiving two courses; 8 percent, 3 courses; 3 percent, 4 courses; and 3 percent, 5 to ten courses of treatment.[23] It was suggested in this study that although the relatively low prevalence and intensity of infection in the Chinese brigades may have been due to a low level of transmission, as was probably the case in the Philippines, it could certainly have been related to the high level of treatment reported in the Chinese population.[24]

---

[22] E. O. Domingo, E. Tiu, P. A. Peters, K. S. Warren, A. A. F. Mahmoud, and H. B. Hauser, "Morbidity in Schistosomiasis japonica in Relation to Intensity of Infection: Study of a Community in Leyte, Philippines," *American Journal of Tropical Medicine and Hygiene* 29 (1980): 858–67.

[23] Warren et al., "Study of Two Rural Brigades."

[24] Ibid.

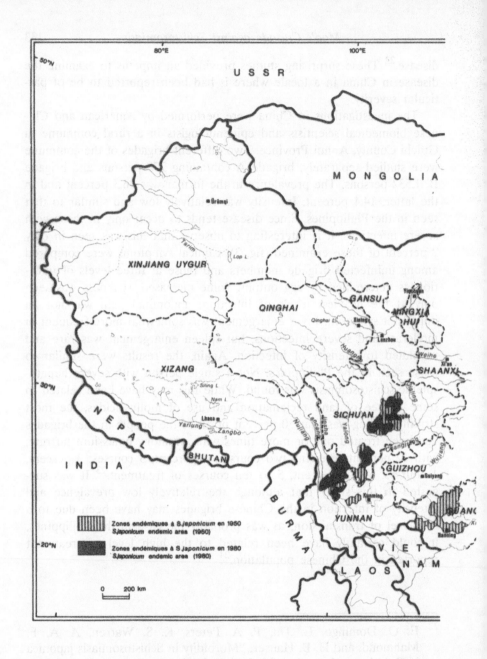

Figure 2. Distribution of Schistosomiasis in China, 1950 and 1980.

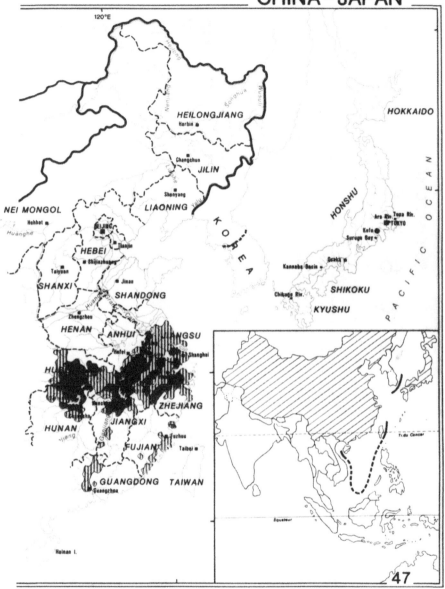

Source: J. P. Doumenge, K. E. Mott, C. Cheung, D. Villenave, O. Chapuis, M. F. Perrin, and G. Reaud-Thomas, *Atlas of the Global Distribution of Schistosomiasis* (Geneva, 1987), 47. Reprinted by permission of the World Health Organization.

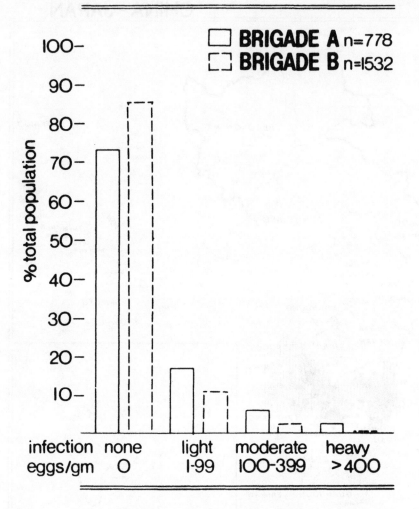

Figure 3. Percentages of Uninfected Subjects and of Subjects with Various Intensities of *S. japonicum* Infection in Brigades of a Commune in Anhui Province.

Source: K. S. Warren, D. L. Su, Z. Y. Xu, H. C. Yuan, P. A. Peters, J. A. Cook, K. E. Mott, and H. B. Houser, "Morbidity in Schistosomiasis japonica in Relation to Intensity of Infection: A Study of Two Rural Brigades," *New England Journal of Medicine* 309 (1983): 1533–39. Reprinted by permission of the New England Journal of Medicine.

While performing the above investigations, researchers had the opportunity to examine the patients in the Guichi County Antischistosomiasis Hospital, which is one of eighteen such hospitals in Anhui Province.[25] The one-hundred-bed hospital had a staff of 180 persons. In 1980, 1,010 hospital admissions for schistosomiasis were recorded. On the surgical service, sixty splenectomies were performed in 1980. On the day of the study, 94 patients, the majority of whom were farmers, were in the hospital. Only 35 percent of these persons were positive for schistosomiasis on stool examination; the arithmetic mean egg count was relatively high at 434 eggs per gram. The most common admission complaint was weakness of at least two weeks' duration; dysentery was the next most common complaint. Hepatosplenomegaly was present in 28 percent of the patients, most of whom did not have eggs in the stool. Most of the patients had been treated multiple times. In comparing the status of the patients in the hospital with that of the people of the commune, the patients were older, and those who were infected had a higher intensity of infection. Liver enlargement was more than twice that seen in the commune and splenomegaly was much more common.

Immediately after brigades A and B were studied, all those infected were treated with a new and highly effective antischistosomal drug, praziquantel. In addition, the lands around brigade A were treated with molluscicides, while those around brigade B were left untreated. When examined one year later, infection prevalence was found to have decreased by 73 percent in brigade A and by 61 percent in brigade B.[26]

The story of schistosomiasis in China has enormous ramifications and raises many questions. First of all, how important was schistosomiasis in comparison with the other diseases and problems plaguing China? What does ten million cases out of six hundred million people mean? (This was 1.6 percent of the population of China when Mao wrote his poem in 1958.) Furthermore, it must be remembered that schistosomes are nonmultiplying helminth parasites (worms). As Faust and Meleney pointed out in 1924, "Since there is no multiplication of this parasite after its entry into the body of the final hosts, the severity of the disease in any one case depends entirely upon the number of cercariae (infective larvae) which penetrate the skin."[27] Since the

---

[25] Personal communication with K. E. Mott.
[26] Personal communication with Su Delong.
[27] Faust and Meleney, "Studies on Schistosomiasis japonica."

distribution of schistosomiasis in most populations tends to be negative binomial or overdispersed, those individuals with moderate to severe infections (and thereby prone to disease) usually make up a very small proportion of the infected population. In the case of the two brigades studied recently, only about 4 percent had moderate infections and 2 percent had severe infections.[28]

In areas where both the prevalence and intensity of schistosomiasis mansoni is high, hepatosplenic disease is relatively common.[29] The Chinese evidence strongly suggests that in localized areas, such as those around Shanghai, prevalence and intensity of *S. japonicum* infection were very high and therefore schistosomiasis was a major problem. Furthermore, when troops who had never before been exposed to schistosomiasis came into the area, many of them probably developed acute schistosomiasis or Katayama fever. All evidence gathered on this disease outside of China suggests that it occurs only with heavy primary infections. These infections, if occurring on a broad enough scale, could have an adverse effect on troop movements. Thus, it appears that there was sufficient reason for Chairman Mao to be concerned about the problem. But was this problem of such importance as to require a massive campaign to eradicate the infection throughout the entire country? A strategy of eradication meant an effort that was immensely greater than that needed to control the infection. An effort to control the disease, which occurs in a relatively small proportion of the infected populations, would have required even lower cost and effort. The primary strategy of snail control required an enormous mobilization of the population. A widespread use of the highly toxic antimony drugs must have resulted in considerable morbidity and mortality, particularly when the three-day regimen was used. How necessary were all the antischistosomiasis hospitals, and was the high rate of splenectomies performed therein (more than ten thousand by 1964)[30] justified, particularly in light of modern findings that splenectomized individuals are far more

---

[28] Warren et al., "Study of Two Rural Brigades."

[29] T. K. A. Siongok, A. A. F. Mahmoud, J. H. Ouma, K. S. Warren, A. S. Muller, A. K. Handa, and H. B. Houser, "Morbidity in Schistosomiasis mansoni in Relation to Intensity of Infection: Study of a Community in Machakos, Kenya," *American Journal of Tropical Medicine and Hygiene* 20 (1971): 273–84.

[30] Cheng, "Schistosomiasis in Mainland China."

susceptible to certain infections, such as pneumonia, typhoid, and malaria?

Now that it has been revealed that schistosomiasis has not been eradicated in China, what is the next step? Is it necessary to continue the massive effort of burying old canals and building new canals, the wholesale broadcasting of poisonous substances like molluscicides, and the use of highly toxic drugs? One thing is certain, and that is that highly toxic antimony-containing drugs are no longer necessary, since praziquantel, a superb new drug, has appeared for the treatment of schistosomiasis japonica. The Chinese Government is now making the drug under the name of pyquiton. Praziquantel will cure most cases of schistosomiasis, with very few side effects, when given orally during the course of a single day. It may very well be that if this drug is used judiciously, it will be possible to control disease and, to a certain extent, infection, until agricultural development, together with land reclamation and the provision of safe water supplies and sanitation, result in schistosomiasis's gradually dying out in China, as it has in Japan.

Nevertheless, the remarkable degree of control of schistosomiasis in China must be considered one of the great global health achievements. The zeal of the Party members in trying to eradicate schistosomiasis apparently overrode the traditional caution of the physicians and the scientists. As David Lampton, a political scientist, has pointed out, schistosomiasis control in the early 1950s in China was under the purview of the Ministry of Public Health and "was influenced by medical and epidemiological experts who urged caution in program implementation." He further notes that the ministry had "little organizational leverage over other ministries whose aid and cooperation was essential, and had insufficiently extensive links with the masses to make widespread mobilization possible." Therefore, in November 1955, the Party's Central Committee created a nine-man subcommittee on schistosomiasis "which was given the authority to coordinate the myriad agencies whose timely cooperation was essential." As the subcommittee grew in strength, it became increasingly antiprofessional and its policies were conceived and executed without the restraining influence of professionals. Lampton noted that while it was necessary to coordinate activities above the ministry level, it was also important to make sure that medical and professional considerations were taken fully into account.[31]

The information related above, much of which depends on the relatively incomplete data now available, suggests several general conclusions. First, the attempt to eradicate schistosomiasis in the whole of China was probably unwarranted on the basis of the tools then and even now available. A far less costly strategy would have involved disease control in most of the infected areas and infection control, including agricultural engineering and mollusciciding, in localized areas of high severity. Second, the remarkable results of the eradication campaign are due more to the personal interest of Chairman Mao and the mobilization of the people rather than to the efforts of health officials and scientific experts. Third, continuing attempts to eradicate schistosomiasis in China are not warranted by the relative importance of the problems and the tools now available. Control of disease can now be achieved relatively simply through the use of the new antischistosomal drugs. Snail surveillance and control should be maintained in the areas where near-eradication has been achieved. If better means of snail control become available or the great promise of biotechnology leads to a vaccine, more intensive efforts will be justified and the great dream of Chairman Mao, "May we ask Mr. Plague—where do you want to go?" may be fulfilled.

---

[31] D. M. Lampton, "Policy Change and China's Antischistosomiasis Program: An Evaluation," *American Journal of Tropical Medicine and Hygiene* 26 (1977): 458–62.

Human Viral Vaccines in China

Scott B. Halstead and Yu Yong-Xin

Much emphasis has been given in the scientific and lay press to the coming "new age of vaccines," when vaccines will be chemically defined products of genetic engineering or of biosynthesis. In the excitement generated by these reports, it is often forgotten that we are still passing through a remarkable period in the history of human immunization. Classical, empirical virology has made available several effective live-attenuated vaccines. These have few or no side effects, can be produced and delivered at low cost, and have given us several authentic public health miracles, specifically, live virus vaccines against smallpox, poliomyelitis, measles, rubella, and mumps. Although smallpox vaccine achieved progressively wider use following its introduction in the late eighteenth century, the genuine success story of this and other viral vaccines is less than fifty years old. Yellow-fever live-attenuated vaccine, developed in 1937, played a significant role in the conquest of yellow fever in the Americas. The mid-1950s saw the introduction of killed, then live, polio vaccines and the virtual elimination of polio in industrialized countries. These successes emboldened the World Health Assembly to plan a global attack on smallpox. Its work has been carried on by the Expanded Program of Immunization (EPI), which envisions the use of vaccines on a sufficiently large scale to eliminate vaccine-preventable causes of childhood morbidity and mortality.

These prefatory remarks serve as a reminder that powerful immunizing agents now are available that can be used to reduce or eliminate important human diseases. The challenge is mass immunization. While the requirement for cold chains and administrative skills has often been emphasized, an underappreciated element in long-range success is the availability of vaccines that are affordable. The production of live virus vaccines in strategically located developing countries would be an important step toward this goal. Local production not only reduces hard currency needs, it also introduces the technology needed to assess immune status and to monitor the safety and efficacy of viral vaccines.

The historical contribution of China to the introduction and use of live virus vaccines is not widely appreciated. The Chinese contribution is important for three reasons. First, vaccine production in China, which came online almost simultaneously with that in Western countries, has effectively removed about 20 percent of the world's population from the virus-susceptible pool. Second, in mastering mass live-virus vaccine production, China surmounted technical and logistical problems as complex as, if not more so than, those faced today by the majority of developing countries. The Chinese experience is an appropriate model for establishing vaccine production centers around the world. And third, Chinese virological research has successfully developed virus strains equal to or better than those used in Western countries. In addition, China has produced a major new live-attenuated vaccine.

In this paper we will review the development in China of poliomyelitis, measles, and experimental Japanese encephalitis vaccines in the context of global immunization strategies.

Polio

Despite the intermediate status of poliomyelitis as a public health problem in China (compared with Western countries), Chinese virologists moved quickly to master the virological developments of the 1950s. In 1949, Enders, Weller, and Robbins reported the successful propagation of poliomyelitis in human tissue cultures.[1] Recovery of polio

---

[1] J. F. Enders, T. H. Weller, and F. C. Robbins, "Cultivation of the Lansing Strain of Poliomyelitis Virus in Cultures of Various Human Embryonic Tissues," *Science* 109 (1949): 85–87.

virus in China was reported from Shanghai in 1957.[2] Sabin announced the successful development of a live-attenuated polio vaccine in 1957; in the United States, initial supplies were available by 1958. Following the lead of Russian advisors, the Chinese opted to proceed with manufacture of live virus vaccines, a process which is technically more difficult than that for killed vaccine production. The first Chinese oral polio vaccine trial was in 1959[3] and large-scale field trials of Chinese-manufactured type 1 and type 2 vaccines were undertaken in May and June 1960.[4] By 1962, Chinese vaccine was packaged in a hollow candy (called "dragee") and stabilized with a room-temperature shelf-life of several months. No similar formulation is available in Western countries.

As evidence of the comparative speed of the Chinese polio vaccine development program, Japanese-manufactured polio vaccine was not released for use until 1961. In the only comparably-sized developing country, India, a locally made vaccine has not yet been marketed. By 1983, only Mexico, Vietnam, and Iran manufactured polio vaccines prepared in-country.[5]

Further evidence of the technical competence of Chinese virology was the development of an indigenous attenuated polio type 3 vaccine. Sabin's type 3 Leon strain proved to be unstable and was the cause of paralytic disease in some countries. Manufacture of this strain was discontinued in China in 1964.[6] In 1969, four polio type 3 strains isolated in China between 1961 and 1963 were subjected to rapid passage in monkey kidney cell cultures at reduced temperature. These were tested

---

[2] F. C. Ku, J. H. Xiao, D. Z. Zhu, et al., "Isolation and Identification of Poliomyelitis Viruses in Shanghai" (in Chinese), *Chinese Journal of Infectious Disease* 1 (1958): 228–31.

[3] F. C. Ku, Y. Zeng, C. S. Mao, et al., "Serological Response in Children under Seven Years of Age to Trivalent Sabin's Live Polio Vaccine," *Chinese Medical Journal*, no. 47 (1961): 423–28.

[4] F. C. Ku, P. J. Chang, Y. L. Cheng, et al., "A Large Scale Trial with Live Polio Virus Prepared in China," *Chinese Medical Journal*, no. 82 (1963): 131–37.

[5] R. H. Henderson, "Expanded Program on Immunization" (personal communication).

[6] F. C. Ku, D. X. Dong, O. S. Jhi, J. T. Niu, and H. H. Yang, "Poliomyelitis in China," *Journal of Infectious Disease*, no. 146 (1982): 552–57.

for ability to grow at high temperature (t marker), and ability to grow at a low pH (D marker). Viruses showing these properties were subjected to repeated plaque purification and tested for neurovirulence in monkeys. The Zhong III-2 strain with reduced monkey neurovirulence produced good immunity in human volunteers following oral administration. In 1971, this polio type-3 vaccine was licensed for use in China.[7] National vaccine dosage requirements are in excess of two hundred million per year. The incidence of poliomyelitis in China has fallen from 3.52 per 100,000 in 1965 to 0.70 in 1980.[8]

Measles

Measles virus was isolated by Enders and co-workers in 1954.[9] As measles is a slow-growing virus with indistinct cytopathic effect, it was some years before improved assay techniques permitted virus attenuation attempts. In the United States, by 1960, serial passage in chick-embryo tissue culture produced a virus of reduced virulence for man, but which in most vaccine recipients produced fever and rash.[10] These unwanted measles-like reactions were reduced by giving vaccine along with measles immune globulin.[11] Higher passage in tissue cultures resulted in the recovery of further attenuated strains of measles and their incorporation into licensed vaccines by 1963.[12]

---

[7] Ibid.

[8] Ibid.

[9] J. F. Enders and T. C. Peebles, "Propagation in Tissue Cultures of Cytopathogenic Agents from Patients with Measles," *Proceedings of the Society of Experimental Biology and Medicine*, no. 86 (1954): 277–86.

[10] S. L. Katz et al., "Studies on an Attenuated Measles Virus Vaccine: General Summary and Evaluation of Results of Vaccination," *New England Journal of Medicine*, no. 263 (1960): 180.

[11] R. Werbel et al., "Administration of Enders' Live Virus Vaccine with Human Immune Globulin," *Journal of the American Medical Association*, no. 180 (1962): 1086.

[12] A. J. F. Schwarz, "Preliminary Tests of a Highly Attenuated Measles Vaccine," *American Journal of Diseases of Childhood*, no. 103 (1962): 386.

Despite the technical difficulties of measles culture *in vitro,* the virus was successfully isolated in China in 1958; virus attenuation attempts began in 1960.[13] At least three attenuated viruses were developed: S-191 (Shanghai), C-47 (Changchun), and P-55 (Beijing), the latter two derived from the Leningrad-4 strain. These vary in reactogenicity from moderate (S-191) to low (C-47 and P-55). Initially, these measles viruses had been recovered in human kidney or human amnion cells and then adapted to chick embryo cells.[14] Attenuation strategy involved inoculation of different lots of chick embryo cells and then testing high and low passaged measles virus in children volunteers. Large-scale tests were carried out on the S-191 strain in 1963–64.

The amount of time it took to derive attenuated measles viruses in China compares favorably with that in Japan, a country with ready access to Western technology. Two measles vaccines were produced in Japan in 1961, one by Okuno et al.,[15] who attenuated the Toyoshima strain by serial passage in chick embryo, and the other by Matumoto and associates, who propagated the Sugiyama strain in calf kidney cells.[16] In 1962, these two strains were shown to be underattenuated in phase I testing in man.[17] In Japan, prior administration of a killed vaccine was used to reduce measles-like symptoms, and from 1966 to 1968 millions of children were immunized with a sequence of killed, then live vaccine.[18] By 1968, reports of atypical measles in children immunized by

---

[13] Xiang Jianzhi and Chen Zhihui, "Measles Vaccine in the People's Republic of China," *Reviews of Infectious Diseases* 5 (1983): 506–10.

[14] Huang Chenxiang, Chia Ping-yi, Chu Fu-t'ang, et al., "Studies on Attenuated Measles Vaccine. I. Clinical and Immunologic Response to Measles Virus Attenuated in Human Amnion Cells," *Chinese Medical Journal,* no. 81 (1962): 9–22.

[15] Y. Okuno, M. Takahashi, K. Toyoshuma, et al., "Studies on the Prophylaxis of Measles with Attenuated Living Virus. III. Inoculation Tests in Man and Monkeys with Chick Embryo Passaged Measles Virus," *Biken Journal,* no. 3 (1960): 115–20.

[16] M. Matumoto, M. Mutai, Y. Saburi, et al., "Live Measles Vaccine: Clinical Trial of a Vaccine Prepared from a Variant of the Sugiyama Strain Adapted to Bovine Kidney Cells," *Japanese Journal of Experimental Medicine,* no. 32 (1962): 433–48.

[17] M. Hiroyama, "Measles Vaccines Used in Japan," *Reviews of Infectious Diseases,* no. 5 (1983): 495–503.

[18] Ibid.

killed vaccine and later infected with wild measles virus resulted in the withdrawal from the market of inactivated vaccines.[19] To further attenuate measles vaccine, the Biken Institute passaged Tanabe strain successively in monkey kidney, human kidney, chick embryo amniotic sac, and chorioallantoic membrane, followed by a variety of clonal selection procedures. This resulted in a strain of reduced reactogenicity (fever in 58.7 percent of vaccinees; rash in 22.3 percent), which compared favorably with the Schwarz vaccine licensed in the United States in 1963. This virus was licensed in Japan in 1971.[20] In 1983, two further attenuated measles strains were still being evaluated for licensing (fever, 23 percent; rash, 13 percent).[21]

Although extensive comparable data on reactogenicity of measles vaccine are not available for China, acceptable measles vaccines have been produced since the mid-1960s. Because killed vaccines were not used, the atypical measles problem was avoided.

Measles vaccines are manufactured in six vaccine production institutes. Those in Beijing and Shanghai each make twenty-five million doses per year. Twenty-five percent of the output of the Beijing Biological Production Institute is freeze-dried. Measles attack rates fell from 900 to 155 per 100,000 population in 1950–65 to less than 1 per l00,000 in the mid- to late 1970s.

Excluding Western nations and China, in 1983 measles vaccine was being manufactured only in Vietnam, South Korea, and Iran.[22]

Japanese Encephalitis Vaccines

Japanese encephalitis (JE) is a zoonotic virus antigenically related to yellow fever and dengue, and transmitted to man by *Culex tritaeniorhynchus,* a rice-paddy breeding mosquito. Japanese encephalitis is widely distributed in China, producing outbreaks in the pre-vaccine era of from

---

[19] Ibid.; and V. A. Fulginiti, J. J. Eller, A. W. Downie, and C. H. Kemp, "Altered Reactivity to Measles Virus: A Typical Measles in Children Previously Inoculated with Killed Virus Vaccine," *Journal of the American Medical Association,* no. 202 (1967): 1075–80.

[20] Hiroyama, "Measles Vaccines Used in Japan."

[21] Ibid.

[22] Henderson, "Expanded Program on Immunization."

fifty to a hundred thousand cases, with fatality rates of 40–60 percent. JE causes encephalitis and death in horses and abortion in swine. JE is the cause of encephalitis in man in all of the rice-growing countries of Asia as far west as India.

Inactivated Virus Vaccine

In 1951–53, workers at the Institute for the Control of Pharmaceutical and Biological Products made pilot lots of killed Japanese encephalitis vaccine prepared in suckling mouse brain and chick-embryo cell cultures. During 1952–57, a 10 percent formalin-activated vaccine was mass produced. A two-dose regimen offered 85 percent protection in man. From 1958 through 1966, a second killed vaccine made in chick embryo was used widely. This vaccine, however, produced local and systemic reactions. In 1967, a killed JE vaccine made in primary hamster-kidney cells (HKC) from the P-3 Beijing JE strain was introduced. In HKC, JE virus replicates to a titer of 10^7 TCID50 per milliliter. The vaccine is stable at 37°C for two months and at 10°C for three years. Currently, more than 48 million children are immunized annually with HKC killed JE vaccine, with 0.5 milliliter doses given twice (one week apart), then six months later, and then at intervals of four, seven, and ten years. Two doses of the HKC vaccine produce a detectable antibody response in 65.8 percent of antibody-negative children and protection against encephalitis in 76–94 percent.

Comparable dates for JE vaccine production in Japan are the introduction of a killed mouse-brain vaccine in 1954[23] and a purified and concentrated mouse-brain vaccine in 1965.[24] Virtually the entire childhood population of Japan has received three or more doses and epidemic JE has ceased to exist.[25] Purified JE vaccine in mouse brain is expensive to manufacture, currently priced at US $8.00 per dose. The requirement for large-scale ultracentrifugation and the need to rear large numbers of laboratory mice have effectively prevented transfer elsewhere of this method of vaccine production.

---

[23] S. Matsuda, "An Epidemiological Study of Japanese B Encephalitis with Special Reference to Effectiveness of Vaccination," *Bulletin of the Institute of Public Health* (Tokyo), no. 11 (1962): 173–90.

[24] K. Takaku, T. Yamashita, T. Osanai, et al., "Japanese Encephalitis Purified Vaccine," *Biken Journal*, no. 11 (1968): 25–39.

[25] Matsuda, "Epidemiological Study of Japanese B Encephalitis."

JE Live-Attenuated Vaccine

Despite the large-scale JE vaccination programs in China, as many as ten thousand cases of JE still occur annually. This may be attributed in part to the poor immunogenicity of killed vaccine and in part to incomplete vaccination of target populations. There is a need for a vaccine which successfully immunizes with a single dose. This describes a live-attenuated vaccine.

Workers at the National Institute for the Control of Pharmaceutical and Biological Products began studies on biological and antigenic variations among JE viruses in 1960. From these studies the parental virus selected for vaccine development was SA14, a viral strain recovered from *C. tritaeniorhynchus* in 1954. Pilot lots of vaccine from this strain had exhibited strong protection when challenged by a panel of wild JE strains isolated in China.

SA14 was passaged serially one hundred times in baby hamster kidney cells. Several small plaque clones were selected which had reduced neurovirulence for mice. On further passage in tissue culture, plaque size and neurovirulence revertants were detected. Further plaque cloning was done on clone 12 virus lineage and an avirulent, stable clone, SA14-5-3, was identified.[26] This virus was passaged subcutaneously in suckling mice six times.[27] The virus was then cloned twice and SA14-14-2 selected.[28]

---

[26] Y. X. Yu, G. Ao, Y. G. Chu, T. Fong, N. J. Huang, L. H. Liu, P. F. Wu, and H. M. Li, "Studies on the Variation of Japanese B Encephalitis Virus. V. The Biological Characteristics of an Attenuated Live-Virus Vaccine Virus Strain," *Acta Microbiologica Sinica* 13 (1973): 16–24.

[27] Y. X. Yu, C. Fang, P. F. Wu, and H. M. Li, "Studies on the Variation of Japanese B Encephalitis. VI. The Changes in Virulence and Immunity after Passaging Subcutaneously in Suckling Mice," *Acta Microbiologica Sinica* 15 (1975): 133–38.

[28] Y. X. Yu, P. F. Wu, J. Ao, L. H. Liu, and H. M. Li, "Selection of a Better Immunogenic and Highly Attenuated Live Vaccine Virus Strain of Japanese B Encephalitis. I. Some Biological Characteristics of the SA14-14-2 Mutant," *Chinese Journal of Microbiology and Immunology* 1 (1981): 77–83.

As shown in figure 1, both the 5-3 and 14-2 strains show desirable attenuation characteristics, i.e., reduced neurovirulence for mice and for monkeys (5-3 strain).

When inoculated in man using comparable infectivity doses, 14-2 showed superior immunogenicity to 5-3 and to killed vaccine (table 1).[29]

JE 5-3 has been given to approximately five million children. In a large-scale trial in Guangdong Province, the incidence of Japanese encephalitis was compared in vaccinated persons and in controls. As shown in table 2, protective efficacy, although not perfect, compares favorably with many vaccines in current use. There is no comparable vaccine in Japan or in Western countries.

Discussion

This paper was stimulated by a visit by Dr. Halstead to the National Institute for the Control of Pharmaceutical and Biological Products in Beijing in September 1983. The trip provided several insights into the Chinese viral vaccine program.

First, despite limited exchange of technical information and personnel, Chinese scientists have demonstrated a consistent ability to rapidly adopt discoveries or techniques developed elsewhere in the world. Evidence that this trend continues was provided by observations that several research teams in the institute were producing monoclonal antibodies to viral reagents.

Second, the Chinese have been particularly adept at scaling up vaccine production. With attenuated measles strains, technical developments described in the United States were duplicated and applied to mass production in China sooner than in the United States.

Third, the fact that the Chinese vaccine program is not as widely recognized as it should be may be due in part to a lack of rigor in field tests of efficacy. Detailed statistics on side reactions, seroconversion

---

[29] J. Ao, Y. X. Yu, Y. S. Tang, B. C. Cui, D. J. Jia, and H. M. Li, "Selection of a Better Immunogenic and Highly Attenuated Live Vaccine Virus Strain of Japanese B Encephalitis. II. Safety and Immunogenicity of Live Japanese B Encephalitis Vaccine SA14-14-2 Observed in Children," *Chinese Journal of Microbiology and Immunology* 3 (1983): 245–47.

Passage history of SA14 Strain	Neurovirulence			Reversion to Virulence	Plaque Size
	$Log_{10}LD_{50}$		deaths/total		
	Mouse		Monkey		
	ic	sc	ic		
SA 14	6.0	3.0	10/10		L
↓					
HKC 28	2.0	0.6	1/4		L, S
↓					
HKC 100	2.0	0	0/4		S
plaque cloning					
1 2 3 ... 9 10 12					
Log_{10} LD_{50} 5.50 4.77 3.56 1.0 1.0 1.0					
plaque cloning 2x					
12-1-7	0	0	0/1	HKC mice*	S
plaque cloning x10					
9-7	0	0	0/16	HKC mice*	S
plaque cloning x2					
2-8 Institute of Virology 5-3	0	0	0/16	HKC mice	S
plaque cloning x2					
14-2	0	0	N.D.	mice	S

*Increased neurovirulence after one passage in tissue culture in mice compared with plaque-derived virus.

Figure 1. Japanese Encephalitis: History of Live Virus Vaccine.

Source: Y. X. Yu, P. F. Wu, J. Ao, L. H. Liu, and H. M. Li, "Selection of a Better Immunogenic and Highly Attnenuated Live Vaccine Virus Strain of Japanese B Encephalitis. I. Some Biological Characteristics of the SA14-14-2 Mutant," *Chinese Journal of Microbiology and Immunology* 1 (1981): 77–83.

Table 1

Neutralizing Antibody Response in Children Following Vaccination with Live and JE Killed Vaccine

Vaccine Strain	Dose log TCD50	No. Vaccinated	No. with Antibody Response	%	Seroconversion Geometric Mean Titer
14-2	6.0	13	12	92.3	29
	5.0	17	12	70.6	10
	4.0	16	10	62.5	10
	3.0	19	5	26.3	5
5-3	6.5	13	8	61.5	5
Killed vaccine	0.5 ml	17	4	23.5	11

Source: J. Ao, Y. X. Yu, Y. S. Tang, B. C. Cui, D. J. Jia, and H. M. Li, "Selection of a Better Immunogenic and Highly Attenuated Live Vaccine Virus Strain of Japanese B Encephalitis. II. Safety and Immunogenecity of Live Japanese B Encephalitis Vaccine SA14-14-2 Observed in Children," *Chinese Journal of Microbiology and Immunology* 3 (1983): 245–47.

Table 2

Protective Efficacy of 5-3 Japanese Encephalitis Live-Attenuated Vaccine: Five-Year Follow-up of Children Vaccinated in 1975

Years	Vaccinated Group		Control Group		Protection Rate (%)
	No. Vaccinated	No. Cases Encephalitis	No. Observed	No. Cases Encephalitis	
1973	205,359	58	26,180	63	88.26
1974	205,301	12	26,117	22	94.06
1975	205,289	8	26,095	7	85.49
1976	205,281	7	26,088	13	92.98
1977	205,274	3	26,075	9	95.77

Source: Ao et al., "Selection of a Better Immunogenic Vaccine (II)."

rates, and protective efficacy, while perhaps available in unpublished reports, are not in the medical literature. Reports of efficacy are often categorized in generalities, such as "good," rather than in detailed, statistically sound data sets. The effect of this imprecision may be to prevent acceptance of the information presented and hence to encourage disbelief in the vast accomplishments of Chinese virology.

Finally, distribution problems in China are unusually complex, due to the size of the country and of the population to be vaccinated. A consistent problem observed by Dr. Halstead was the lack of monitoring of the immunogenicity of vaccines actually administered in the field. This has led to the assumption in China that vaccines are often inactive when administered and to a policy of redundancy in immunization schedules. While this seems to be an effective solution, improvements in the cold chain and a vaccine stability monitoring system should be more cost-effective in the long run.

Despite these problems, developing countries have much to learn from the Chinese model. The viral laboratory facilities used in China during the past thirty years differed little in quality from those in other developing countries. However, the Chinese educational and managerial system seems to encourage innovation and pragmatism. The facilities of the Beijing Biological Production Institute, while spartan, were filled with engineering innovations, each using available materials but demonstrating clever solutions to the problems of large-scale production of vaccines.

China presents interesting and unique opportunities for the conduct of large-scale trials of vaccine efficacy. Unpublished data from the National Institute for the Control of Pharmaceutical and Biological Products suggest that a single dose of killed JE vaccine may protect against encephalitis. If true, this could lead to a cost-effective method to prevent Japanese encephalitis. HKC vaccine is inexpensive to manufacture.

In India, Nepal, Thailand, and Vietnam, JE outbreaks have increased in size progressively over the past twenty years. Because of the high ratio of asymptomatic infections to encephalitis cases, about 50:1, each of these countries will need to immunize millions of persons yearly to effect a significant reduction in cases. An inexpensive vaccine is needed urgently. A careful evaluation of strategies for the use of one or more doses of killed vaccine in China would make a valuable contribution to the development of vaccine programs in tropical Asia. The other exciting possibility is the Chinese live virus vaccine. The manufacturing process is simple; no purification or concentration is required. Preparation from hamster-kidney cells is relatively simple. Well-designed efficacy trials are needed; these should include the measurement of the actual dose of vaccine virus used in field trials.

The Chinese exhibit what in the West is called the entrepreneurial spirit. In many developing countries, a comparable spirit is needed in the public health sector.

Genetics in Postwar China

James F. Crow

Earlier in this volume, Laurence Schneider emphasized the central role of T'an Chia-chen (C. C. Tan, or Tan Jiazhen in *pinyin* romanization) in Chinese genetics. Tan's story epitomizes the past half-century. I can think of no better way to begin my paper than with a discussion of his work.[1]

A recent selected bibliography of Tan's publications lists twenty-seven articles published between 1932 and 1949, all substantial papers based on original research. From 1950 to 1963 there was none; this period is labeled in the bibliography as "Dominance by Lysenkoism." In 1963 and 1964 there were four articles, then no more until 1979. The period from 1966 to 1976 is listed as "Dominance by Cultural Revolution." Since 1979 Tan has published fourteen articles. He remains active as a writer,

---

[1] It is a pleasure to acknowledge my great indebtedness to C. C. Tan. I have relied heavily on his lectures and on personal discussions with him. My one-month lecture tour in China was supported by the Committee on Scholarly Communication with the People's Republic of China and Academia Sinica. I am most thankful to Hu Han for his help as my principal host in China and to him and many others for taking a great deal of time to tell me about their institutions, their work, and their experiences during and since the Cultural Revolution. This is paper number 2744 from the Laboratory of Genetics, University of Wisconsin.

and even more as a catalyst and director of the Genetics Institute at Fudan University in Shanghai. But for two periods totaling more than twenty-five years, he did no published research.

As this example illustrates so vividly, Chinese genetics was doubly plagued. It suffered greatly during the "ten long years" of the Cultural Revolution. In addition, the previous fifteen years were dominated by the ingenuous, but pernicious, Russian genetics of the time. Instead of a ten-year eclipse, genetics had twenty-five years in which progress was essentially nil. Chinese genetics is still in the beginning stages of a difficult catch-up process.

My discussion will take two points of departure. The first is Tan Jiazhen and his work. The second is the Institute of Genetics in Beijing, a large research institute. I regret that this article is not based on a scholarly or systematic study; my knowledge and experience are too limited. Rather, it is a personal account, based on impressions gathered during a one-month trip in 1983 and discussions with Dr. Tan.

Tan Jiazhen

My acquaintance with Tan Jiazhen began in 1946. At that time I was at the University of Texas, where he gave a talk on color patterns in ladybird beetles. He described "mosaic dominance," whereby the color in each body part reflects the dominance of the genes expressed in that specific tissue. I was aware of similar results involving bristle patterns in the vinegar fly, *Drosophila*, which some Russian geneticists had interpreted as reflecting a corresponding subdivision of the gene. This preformationist interpretation did not appeal to either of us, but it made for an interesting discussion. We also shared, and still do, an interest in *Drosophila* population genetics. Furthermore, we both have a vivid memory of his attempt to get authentic Chinese food in a Texas Chinese restaurant by speaking to the waiters in their native language, only to find that no one there could understand Chinese. Alas, the initial friendship was very brief; I didn't see him again for more than thirty years.

Professor Schneider has mentioned the great influence of Li Ju-ch'i (Li Ruqi in *pinyin*) in bringing *Drosophila* genetics to China. Li worked in T. H. Morgan's lab at Columbia University and published several papers with C. B. Bridges in the 1920s. Tan was one of Li's students, and they worked together on ladybird beetles. Li sent one of Tan's papers to Morgan, by that time at the California Institute of Technology, who passed it on to Th. Dobzhansky. Dobzhansky had recently come from Russia to work with the Morgan group and especially with A. H. Sturtevant. Dobzhansky had worked on ladybird beetles in Russia and immediately it was arranged for Tan to work at Cal Tech. He was there from 1933 through 1939 and for a shorter visit in 1945–46.

In the 1930s Dobzhansky and Sturtevant were developing the genetics of *Drosophila pseudoobscura* and its relatives. This was the beginning of the heyday of evolutionary studies in *Drosophila* species. Tan immediately became expert in the study of giant salivary-gland chromosomes in *Drosophila* and did a pioneer cytological analysis of two closely related species. He also showed the homology of some eyecolor genes in *D. pseudoobscura* and *D. melanogaster*, two quite different species that cannot be crossed. This was done by the delicate operation of transplanting eye-primordia from one species to the other. Further, Tan and Sturtevant showed the correspondence in chromosomal position of similar genes in these species. This helped discredit the views of Richard Goldschmidt, according to whom species differences were more than the accumulation of smaller, intraspecies differences, and instead required "macromutations." (The Goldschmidtian idea has recently been revived by some paleontologists, unnecessarily in my view, to explain discontinuities in the fossil record.) This was a period of great productivity by the Cal Tech laboratory and by Tan, and he rode the crest of the wave of popularity of *Drosophila* studies on the origin of species.

After returning to China, Tan continued his work on *Drosophila* and ladybirds. The emphasis was on the study of natural populations and on the mechanisms of species formation. Among other things, Tan discovered regular seasonal changes in color pattern, which he explained by the differential protection of the different color patterns in different seasons. As in the United States, the study of the genetics of natural populations became a central part of genetics.

Then came the period of dominance by the Russian charlatan, Lysenko. Tan wrote nothing during this period. After it again became possible to study Mendelian genetics, Tan returned to the subject. But work on ladybirds and *Drosophila* was too impractical for the political climate of the time. Instead, he studied the genetic effects of radiation in monkeys. This was in the early 1960s, and his research reflected the worldwide interest in this subject at that time.

Following this brief period came the second eclipse of genetics, the Cultural Revolution. Almost nothing could be done for a decade. After this Tan again returned to genetic research. He has continued on his own and with students to study ladybird beetles and *Drosophila*. He has also done some work on cabbage species and on microbial tests for mutagenesis. He and a student are involved in the latest *Drosophila* hot subject, transposable genes. Much of his time during a recent visit to Wisconsin was spent in catching up on American advances in this fast-moving area.

Tan has also been active in administrative work. He has been vice president of Fudan University, where he now directs the Genetics Institute. He frequently gives talks on the history and status of genetics in China and was the organizer of an international conference on chemical mutagens held in Shanghai in 1983. He has provided inspiration and leadership for a new community of young Chinese geneticists.

In discussing the past half-century of genetics in China, I can do no better than to follow an outline used by Tan in recent seminar talks.

The Past Half-Century: Five Periods

1920–1949: The American Influence

As Professor Schneider has emphasized, Chinese genetics during this period was strongly influenced by two American universities, Columbia and Cornell. The two most important organisms for genetic research at that time were *Drosophila* and maize, and these two universities were the respective leaders.

Drosophila genetics was started by T. H. Morgan at Columbia University. Columbia also had the leading American cytologist, E. B. Wilson. Morgan was insightful, and undoubtedly lucky, in two ways. He chose an organism that turned out to be admirably suited for genetic research (no one could have predicted that *Drosophila* would have giant salivary-gland chromosomes), and he picked three remarkably gifted students: Calvin Bridges, A. H. Sturtevant, and H. J. Muller. Thanks to these three, soon to be followed by many others, the subject grew explosively. Morgan and Muller later won Nobel prizes. Morgan was eventually left behind and returned to his first love, embryology, but he remained their respected leader. By the time of Tan's visit the Columbia group had moved to Cal Tech. The Morgan influence was felt in China through Ch'en Tze-ying (Chen Zinying in *pinyin*) and, as I have mentioned, Li Ruqi and Tan Jiazhen.

The other great genetic influence was Cornell. R. A. Emerson, although his own work on maize did not gain wide attention (he was overshadowed by E. M. East at Harvard and L. A. Stadler at Missouri), had Morgan's gift for attracting outstanding students. At one time he had George Beadle, Barbara McClintock, Charles Burnham, and Marcus Rhoades, the first two Nobel Prize winners and the others of comparable eminence. As Schneider has said, several Cornell graduates were influential in the development of genetics and plant breeding in China.

At least two geneticists from Columbia and Cornell were still active in 1983. At Beijing University Li Ruqi was an alert and well-preserved 89. While a student at Purdue he had played football. Until recently he had bicycled to work and taught genetics courses. Zhou Chengyao (C. Y. Zhou) was 81 and a retired professor of agronomy at Zhejiang Agricultural University in Hangzhou. He had studied at Cornell and was associated with the Emerson group of the late 1920s.

Cornell was also strong in agronomy and plant breeding. This had three consequences for China. One was high-quality plant genetics in the Emerson tradition. The second was sophisticated plant breeding. The third might have been important, but was not. Li Zhongjun (Li Chung-chün, or C. C. Li) studied agronomy at Cornell and became proficient in statistics and population genetics. He returned to Beijing where he joined the faculty of the College of Agriculture at Beijing University (then known as National Peking University). While there he wrote a modern classic, *An Introduction to Population Genetics*. It presented for the first time a readable account of the mathematical theory of population genetics as developed by R. A. Fisher, J. B. S. Haldane, and Sewall Wright. It was published in Peking in 1948 and was available in English soon after. Li's gift for simplifying and explaining, and his sympathetic appreciation of what students can be expected to understand and not to understand, were apparent on every page. I read it avidly from beginning to end, as have many others since that time. But Li did not stay in China and mathematical genetics did not develop there. I shall return to his departure later.

1950–1957: The Michurin-Lysenko Period

I. V. Michurin was a Russian horticulturalist, somewhat akin to Luther Burbank in this country. Both were widely known as breeders, but their theoretical ideas were at best confused. Michurin was a fruit grower with considerable success in grafting. Lysenko took up the Michurin view and, being strongly supported by Stalin, attained great political power in the late 1930s that continued through the 1950s. Lysenko held that genes are nonexistent hypothetical constructs, that chromosomes are irrelevant to heredity, and that the effects of environmental adaptations are inherited. Given the strong Russian influence in the 1950s, it is natural that Lysenkoism became the official genetics of China.

The strong appeal of Michurin-Lysenko genetics is not hard to understand. It bypasses the necessity to understand the complex behavior of chromosomes during meiosis. The intellectual rigor of chromosome mapping and the difficult techniques for cytological study are not needed. There is no need to understand statistics. This enticed the ignorant and the lazy. Most importantly, it promised immediate results; there was no need to wait several generations for Mendelian selection to work.

Early Russian plant genetics had been led by Nicolai Vavilov, whose program was to gather, from all over the world, wild relatives of agricultural plants. By crossing these with cultivated varieties and selecting among the segregating progenies, he hoped to incorporate useful properties, such as winter hardiness and resistance to fungal and insect diseases, from the wild relatives. Vavilov was also interested in the more academic question of the origin of plant species and their geographical distribution. His program was thoroughly sound—indeed, such studies are the basis for much of what goes on currently in agricultural experiment stations throughout the world. But this is a program of patient, hard work. The results come slowly. It is no wonder that Lysenko's rash promises were preferred. The lure of easy, immediate results is hard for political leaders to resist.

During the period of Russian-Chinese friendship, Lysenkoism moved rapidly through China. Large numbers of Russians spread the gospel, and Chinese geneticists went to Russia for study. Michurin-Lysenko genetics soon became *de rigeur* and essentially all genetics taught during this period was of this school. Classically trained geneticists, if they continued teaching or research, simply did something else. One that I know changed to plant physiology and did research in this area. Tan Jiazhen continued to teach, but taught evolution instead of genetics. The course was devoted mainly to paleontology and the evidences for evolution, rather than its mechanisms. Genetics research in the Western tradition was essentially nil.

I do not want to imply that all Lysenkoist genetics was a total failure in a practical sense. However one thinks heredity works, it is good practice to breed from the best plants and animals. Burbank achieved some astonishing results by having an eye for good, novel variants and by growing enormous numbers from which to select such plants. Presumably, the same thing happened in China. But not all plant breeding is so simple. The greatest success story in American plant breeding, hybrid corn, makes no sense in Lysenkoist terms. Also, the Lysenkoist rejection of statistical methods meant that the most efficient field designs, developed by the "reactionary" British statistician and geneticist, R. A. Fisher, were not employed in the routine testing of new varieties.

Let me now return to the story of Li Zhongjun. Being in Beijing, he was near the center of anti-Mendelian political activity, and he was very badly treated. His wife, who had grown up in the United States, was especially unhappy. He finally managed to get out of China through Hong Kong, at considerable risk to himself and to some loyal friends. In 1950 H. J. Muller, who was a great humanitarian and used his wide influence to help needy scientists, was trying to help Li get into the United States. He knew that I thought very highly of Li's book and, as president of the American Society for Human Genetics, asked if I would write a review for the society's journal, which was published in 1950. Li visited Wisconsin shortly thereafter. Our mutual interest in population genetics, as well as our shared difficulty in understanding the work of Wright, Fisher, and Haldane, guaranteed an immediate friendship. Muller found Li a position at the University of Pittsburgh, something that Pittsburgh has never had occasion to regret.

Li's departure was a great loss for Chinese genetics. Theoretical population genetics hardly exists in China. Things might have been different had Li remained and had it been possible for Mendelism to thrive. On my recent visit there, several people who were in Li's classes remembered the greatness of his teaching some thirty-five years earlier. Visitors to China are impressed with the advanced mathematics taught in school. I am sure there would have been many mathematically gifted students who would have been receptive to Li's teaching and eager to carry the subject forward.

One small, but poignant, aside. Li Zhongjun's book is dedicated "to the memory of Jeff." Only this year did I learn who Jeff was, from a recent biographical essay by Li's student Eliot Spiess.[2] Jeffrey was an infant son who died during the hardship of the last days in China.

1958–1966: Coexistence of Two Genetics

The death of Stalin in 1953 brought some softening of Lysenkoism in Russia, although Lysenko's influence continued for some time. In China the famous Mao proclamation "Let one hundred flowers blossom and a hundred schools of thought contend" was used to advantage. Tan was himself influential in persuading Chairman Mao to let Mendelism contend as one school of thought.

Starting in 1958 Mendelism could again be taught. But Michurin genetics was official; for some time Mendelism remained unofficial. High school courses emphasized Michurinism. It is not hard to imagine the confusion of university students who studied courses of both varieties.

In the early 1960s genetic research was again permitted, but it retained another Russian flavor—great emphasis on the immediately practical. Tan could not study his ladybirds and *Drosophila* and instead studied the effects of radiation on spermatogenesis in *Rhesus* monkeys. This was done in cooperation with N. P. Dubinin, a Mendelian geneticist who was now back in favor in Russia. Other Chinese geneticists turned to plant breeding, to the genetics of human metabolic diseases, to bacterial drug resistance, to human chromosome anomalies, and to bacterial systems for detecting mutagens. Many of these were highly worthwhile ventures, but basic genetics research remained virtually nonexistent.

Yet, Mendelian genetics could again be studied and gradually became predominant. There was every reason to think that genetics would soon flourish; but this hope was dashed by the Cultural Revolution.

---

[2] Eliot Spiess, "Chung Chun Li, Courageous Scholar of Population Genetics, Human Genetics, and Biostatistics: A Living Essay," *American Journal of Medical Genetics* 16 (1983): 603–30.

1967–1976: The Cultural Revolution

This was the period when all science, indeed, all intellectual activity, suffered. Whole departments were consigned to remote areas for "rural reeducation." One group of maize geneticists was sent to grow corn in an area that had much too short a growing season. Several crops were lost by early frost. The whole experience was a fiasco, to say nothing of the personal hardships. What genetics remained had to be aimed toward immediate practical gains, and the Lysenkoist influence still persisted in high places.

It took cleverness and political skill to get along in this period. Tan Jiazhen fared better than many. I will mention one example of his fancy footwork. In the early 1970s, one politically acceptable scientist reported that by injecting a vital dye into cotton plants he had produced colored cotton in subsequent generations, an obvious boon for cotton breeders. Tan went to visit this man and learn the technique, and he tried to repeat the experiments, with no success. Later he found himself included as coauthor of a manuscript reporting these results. Not wishing to destroy his reputation among legitimate geneticists, and yet not wishing to get into political trouble, he wrote that he had indeed studied the techniques. Yet his inability to repeat the experiments showed that he had not sufficiently mastered these techniques and, therefore, did not deserve to have his name on the paper. His name was removed and the paper duly appeared.

1976 and after: Reemergence of Mendelian Genetics

With the end of the Cultural Revolution in 1976, Mendelian genetics again had its chance to grow. The early growth was slow. Ultraleftism still prevailed and most influential geneticists were Russian-trained or at least Lysenkoist in their outlook. Serious growth of genetics began about 1978. Temin, reporting on a trip made in 1977, remarked about the contrast between genetics in China, which played no significant role in biological research and existed mainly in the form of plant and animal breeding, and genetics in the United States, where it is extensively studied and is a central discipline in biology.[3]

Since that time the growth of genetic activity has been very rapid. The number of highly competent geneticists is still small but growing at

---

[3] H. M. Temin, "Basic Biomedical Research," in *Science in Contemporary China*, ed. L. A. Orleans (Stanford, Calif., 1980), 255–67.

a rapid rate as more and more young people are educated in the United States, Japan, and Western Europe.

The Institute of Genetics in Beijing

Most of my time in China was spent at Academia Sinica's Institute of Genetics in Beijing. This institute reflects both the old and the new in recent Chinese genetics, and the rapidity of the change from one to the other. It was founded in 1959 and by 1976 had a staff of 350. In the early years it was Russian-dominated. Until recently it has had no American- or Western European-trained scientists on its staff. Hu Han, the director, was trained in Russia, as were several others.[4]

The institute's 1976 report lists one staff member as the representative of the poor and lower peasants. The "open doorway" called for peasants and workers to join the research work. At the time of a visit by a group of Americans in 1976 there were eighteen peasants in the institute. Each year about one-third of the institute staff went into the field for two to three months. There were also cooperative ventures with hospitals. It was reported that sorghum was greatly improved by the help of peasants' experience. A great deal of publicity was generated by a successful egg transplant in sheep.

Reports in Chinese publications in those days reflected the remaining Lysenko influence. A successful cross between corn and rice was reported by peasants and scientific workers in Jilin Province. (Lysenko had reported similar wide crosses between species with incompatible chromosomes; not believing in the importance of chromosomes, he naturally didn't look at them.) Tail-fin characters from carp were reportedly transferred to goldfish by RNA injections. More spectacularly, characters were transferred from salamanders to goldfish. Needless to say, reverse transcriptase (an enzyme that makes DNA from an RNA molecule), discovered by Temin and Baltimore, was seized on in some circles as evidence for Lamarckian inheritance. A review of Jacques Monod's book, *Chance and Necessity,* was criticized for its assumption that mutations were random.

---

[4] My source for much of this section was the clipping and pamphlet file of the Committee on Scholarly Communication with the People's Republic of China, National Academy of Sciences, Washington, D.C.

A visitor to China today hears none of this. I am told that there are still Lamarckians among the older geneticists in China, those trained during the Russian period. Yet nobody espoused Lysenkoism during my visit at the Institute of Genetics or anywhere else among the places I visited. The discussions were always Mendelian in outlook. Those with whom I spoke were eager to discuss the latest genetic advances in the United States and to tell me of their current research interests. My lectures on molecular evolution and transposable genes were discussed enthusiastically by the audience, during the lecture and in private discussions later. Western textbooks were available, although in short supply. I was pleased to find my own textbook being used.

It is interesting and instructive to look briefly at the present organization and research at the Institute of Genetics. The main departments are: molecular genetics; human and animal genetics; plant somatic cell genetics; cytoplasmic genetics; remote hybrids in plants; applied genetics; and technical support. The emphasis on somatic cell genetics in plants is evident here, as elsewhere in China. It may have been Lysenkoist in its antecedents, but the current work is very modern. Several of the workers at the institute had recently studied in the United States and in England.

Another possible vestige of a Lysenkoist past is the emphasis on "wide crosses." But instead of the ridiculous claims seen in newspaper reports a decade earlier, one now finds solid studies of transfer of useful genes from one species to another, closely related one. Skillful manipulation of haploids and polyploids is commonplace. I found intense interest in learning about techniques of using meiotic mutants to control the amount of meiotic recombination—a bag of tricks that has been of great value in potato breeding here in the States. I was told that growing potatoes from seeds is well developed in northern China. This has enormous practical advantages; tubers are an inefficient means of propagation and are carriers of diseases.

The areas of research that are being emphasized at the Institute of Genetics are molecular genetics and plant somatic cell genetics. There is active research in gene regulation, ribosomes, and chloroplast DNA. About forty people were working on cell genetics. Anther culture was being applied to wheat, corn, rye, rice, and rubber trees; this should be advantageous as a way of getting homozygous lines and thereby achieving better control over the genotype. Protoplast fusion was being used for *Nicotiana* (as elsewhere), but there were active experiments in petunia, barley, and sugar cane. There is an active experimental program involving crosses with wild relatives of domesticated plants. A program in human population genetics was engaged in measuring blood group and isozyme frequencies in many parts of China. The large number of minority groups and the ancient records makes human population genetics especially interesting in China.

The *Annual Report* published in 1982 includes reports on about a hundred projects.[5] Seventy-one papers were published by institute staff members in 1981. During the year there were fifteen foreign visitors from the United States, Germany, and Japan, and each gave one or more lectures. Some of the topics were: basic quantitative genetics; restriction endonucleases; human biochemical genetics; laser instrumentation and microsurgery; genetic engineering; protein secretion through membranes; molecular genetics of yeast; and mutants in cell lines of the Chinese hamster. There is no absence of contact with science from the rest of the world.

Given the twenty-five-year period of domination by Lysenkoism and the Cultural Revolution, the adjustment to modern genetics by this institution is remarkable. I trust the same thing has happened elsewhere in China.

Some General Remarks

China's greatest need for which genetics can supply solutions is to increase the food supply. Although China has an enormous land mass, only a small fraction is arable. From an amount of farm land comparable to that of the United States must come food for a population five times as large. I leave to others a discussion of how successful the birth limitation program is likely to be.

J. R. Harlan believes that good, old-fashioned plant breeding is needed.[6] I have no reason to question this. Increased yield has been one of the biggest success stories in the West. A great deal can be done by crossing and selecting among varieties. Organized yield tests over

---

[5] Hu Han and Shao Qiquan, eds., *Annual Report of the Institute of Genetics*, Academia Sinica, 1981 (Beijing, 1982), 196 and x.

[6] J. R. Harlan, "Plant Breeding and Genetics," in *Science in Contemporary China*, ed. Orleans, 295–312.

wide areas with good statistical control are taken for granted in the West and are increasingly practiced in China. Part of good plant breeding is simply knowledge of the plant, and Harlan believes that farmers can play a useful role in field trials and selection.

I am told that at present agricultural research is not attracting the best students; they are going in for the more highly publicized areas of molecular biology and genetic engineering. For all the promise that genetic engineering holds, it would be a tragedy if the established methods of plant and animal breeding are not used to the fullest. Yet, I did see a number of good programs of plant genetics and plant breeding, which involve some very good people—alert and aware of what is happening throughout the world.[7] Many had been trained in the best Western universities. Yet the number, relative to the total population size, is very small by the standards we are used to, as is the case in other sciences.[8]

There is enormous interest in molecular genetics and genetic engineering, and the titles of papers in current journals are quite similar to those in the United States, although often not quite up to date. In my discussions I found several excellent, well-trained molecular biologists.

They had two kinds of frustrations. One was the lack of a critical mass of scientists similarly interested and a corps of skilled and well-trained assistants; and some complained of their isolation and heavy teaching or other duties. The second major complaint was of inadequate equipment and supplies. I met one young man who had done a beautiful piece of research in the United States and who wanted to continue along similar lines. The lack of equipment and supplies, and especially

---

[7] The educational and research institutions visited, some very briefly, are: Institute of Genetics, Beijing; Beijing Agricultural University; Institute of Developmental Biology, Beijing; Beijing University; Northwest University, Xi'an; Nanjing Agricultural University; Jiangsu Academy of Agricultural Sciences, Nanjing; Jiangsu Institute of Botany, Nanjing; Zhejiang Agricultural University, Hangzhou; Genetics Institute, Fudan University, Shanghai; and Institute of Plant Physiology, Shanghai.

[8] L. A. Orleans, "The Training and Utilization of Scientific and Engineering Manpower in the People's Republic of China," *Science and Technology in the People's Republic of China Background Study No. 5* (Washington, D.C., 1983), 1–71.

of a battery of restriction enzymes, was keeping him from working at full capacity.

The teaching faculties of many universities have little time and almost no facilities for experimental research. It is no wonder that so many biologists have turned to taxonomy in the past. One student asked me to suggest some mathematical problems in population genetics on which he could work alone with pencil and paper, since he expected to be in a place where little else would be available.

The central questions about Chinese genetics, it seems to me, are two. First, can research and education in basic genetics catch up with the United States, Europe, and Japan? And second, to what extent can applied genetics proceed without a strong basic research underpinning?

As for the first question, it is clear that catching up in genetics is harder than in other sciences because of the twin setbacks of Cultural Revolution and Lysenkoism. This means that there is a large age gap among trained geneticists. There are those who were trained in an earlier period and who are now mostly of retirement age; and there are the young, just being trained. The numbers seem terribly deficient to one used to seeing the plethora of geneticists thriving in the United States. There is one important mitigating factor, however. Much of molecular genetics depends more on knowledge of chemistry than of traditional genetics, and chemistry suffered no setback during the Lysenko period. As I mentioned earlier, the level of mathematical training in China is very high. I saw on public television lectures in differential equations and advanced calculus that would be upper division university courses in the States. A Chinese student at Wisconsin told me how surprised he had been to find the low level of mathematics expected in his engineering courses. To the extent that mathematics and the physical sciences can contribute to the molecular genetics of the future, I would expect Chinese scientists to be among the leaders. Yet the problems of numbers and critical mass remain, and genetic knowledge in the West will not stand still while China catches up.

As to the second question, it is clear that a great deal can be done in practical agricultural research without using the latest molecular techniques. It is also clear that an alert, educated corps of workers can utilize new techniques developed elsewhere. In plant and animal breeding, detailed knowledge of the organism is often as important as knowledge of genetics, cytology, and molecular biology. Statistical techniques and experimental design are, I think, coming to be widely understood in experiment stations. How important a strong indigenous basic program is remains unclear. As Temin said, "The next few years will provide an exacting test of the proposition that a community rich in basic scientists is needed as a basis for successful applied science."[9]

In any case, Mendelian genetics is again established in China and is expanding at a rapid rate from its very small base. Lysenko is being forgotten. Mendel is a household word and is taken seriously, as the following conversation illustrates. A student in China asked me whether it was true that Fisher had shown Mendel's data to be "doctored." I replied that modern statistical analysis had indeed shown Mendel's numbers to be closer to the true ratios than could reasonably be expected. He then said, "Why, then, does Mendel's name still appear in textbooks? I should think he would have been denounced and his name expunged." How would you have answered him?

---

[9] Temin, "Basic Biomedical Research."

Agriculture and Plant Protection in China to 1980

Robert L. Metcalf

China has been continuously inhabited by the Chinese since the Paleolithic period (500,000 B.C.), when Peking man, *Homo erectus pekinensis*, hunted large animals, butchered them with stone tools, and cooked them over open fires; yet as much as three-quarters of his diet may have been of vegetable origin. True modern man, *Homo sapiens*, was abundant in China during the Neolithic period (30,000 B.C.), and these ancestors of the Han practiced burial of their dead and had extensive trade with their neighbors.[1] The Han are said to have occupied the Huanghe (Yellow River) valley from time immemorial. Archaeological records leave some uncertainty as to which specific area of modern China was the cradle of agriculture. Was it nurtured in the fertile alluvial plain of the Huanghe, with its abundant water, or as dry farming in the semi-arid loess plateau near the junction of the Huanghe and Weihe? Millet (*Setaria italicum*) was apparently the first domesticated food plant, dating from at least 4000 B.C., and rice (*Oryza sativa*) has been cultivated in North China since 3000 B.C.[2] Over the intervening

---

[1] K. Buchanan, C. P. FitzGerald, and C. A. Ronan, *China: The Land and the People* (New York, 1980).
[2] S. Wortman, "Agriculture in China," *Scientific American* 232, no. 6 (1975): 13–15; and R. C. Hsu, *Food for One Billion: China's Agriculture since 1949* (Boulder, Colo., 1980).

millenia the Chinese succeeded in domesticating many important crop plants that are considered essential today, including soybean, yam, taro, tea, ginger, apple, apricot, peach, and cucumber (see table 1).[3]

The Han employed crop fertilization with human and animal manure from a very early date, and large-scale irrigation appeared by 300 B.C.[4] Trade along the Silk Route offered opportunities for the introduction of new crop plants, domestic animals, and advanced technologies; these many elements were integrated through a very favorable physical environment into a collective society possessing the abundant manpower needed to begin the enormous scale of terracing and irrigation that characterizes central China today. From this beginning, in the last millenium B.C., the ability to control water distribution and to curb flooding made it possible to develop the eastward lowlands of the North China plain.[5] This unrivaled perspective—six thousand years of continuous development of an agricultural industry that has met virtually every demand placed upon it, and has made China self-sufficient in food supply even with a population growth from an estimated eighty million in A.D. 1400 to more than one billion today—provides us with one of history's most important lessons.[6]

Agricultural Overview

Geography and Climatology

The climate of Manchuria from Harbin south to Shenyang has warm summers and long cold winters, with a growing season of about five months. The principal crops are wheat, corn, millet, soybean, sugar beet, and sorghum. North China, from Beijing south to Xi'an, has hot humid summers and cold dry winters, with a growing season of eight months.

---

[3] J. Harlan, "Short List of Chinese Plant Domesticates" (personal communication, 1984); C. B. Heiser, Jr., *Seed to Civlization: The Story of Food*, 2d ed. (San Francisco, 1981); and Loris Milne and Margery Milne, *Living Plants of the World* (New York, 1975).

[4] Hsu, *Food for One Billion*.

[5] Buchanan et al., *China: The Land and the People*.

[6] D. H. Perkins, *Agricultural Development in China 1368–1968* (Chicago, 1969); and Food and Agricultural Organization (FAO), *FAO Production Yearbook*, vol. 35, 1981 (Rome, 1982).

Table 1

Important Crop Plants Domesticated In China

Cereals
Rice — *Oryza sativa*
Millet — *Setaria italicum*

Pulses
Soybean — *Glycine max*
Velvet bean — *Stizolobium hassjo*

Root Crops
Turnip — *Brassica rapa*
Yam — *Dioscorea esculenta*
Lotus — *Nelubium speciosum*
Taro — *Colocasia esculenta*

Spices, Beverages
Ginseng — *Aralia quinquifolia*
Tea — *Camellia sinensis*
Cinnamon — *Cinnamomum cassia*
Ginger — *Zingiber officinale*

Fruits
Chinese hickories — *Carya spp.*
Chinese chestnut — *Castanea henryi*
Japanese persimmon — *Diospyrus kaki*
Kumquat — *Fortunella japonica*
Walnut — *Juglans regia*
Litchi — *Litchi chinensis*
Apple — *Malus asiaticus*
Apricot — *Prunus armeniaca*
Peach — *Prunus persica*
Water chestnut — *Trapa natans*

Vegetables
Garlic — *Allium sativum*
Winter melon — *Benincasa hispida*
Chinese cabbage — *Brassica sinensis*
Cucumber — *Cucumis sativus*

Sources: J. Harlan, "Short List of Chinese Plant Domesticates" (personal communication, 1984); C. B. Heiser, Jr., *Seed to Civilization: The Story of Food*, 2nd ed. (San Francisco, 1981); Loris Milne and Margery Milne, *Living Plants of the World* (New York, 1975).

North China's principal crops are winter wheat, soybean, sorghum, corn, millet, and cotton. Central China, from the lower Yangzi valley south to Changsha, has hot moist summers and cold dry winters, and a growing season of nine months; the principal crops are winter wheat, rice, barley, and cotton. Sichuan has hot moist summers and warm moist winters and a growing season of eleven months, producing rice, yams, wheat, corn, sugar cane, and cotton. South China has hot moist summers and warm dry winters, with a year-round growing season, where two rice crops are regularly harvested along with abundant sugar cane and tropical fruits. Rainfall north of the Huanghe is less than twenty inches per year; south of the Yangzi it is more than fifty inches annually.[7]

The extent and variability of China's agricultural regions can perhaps best be appreciated by superimposing a map of China on that of North America. Harbin lies near Ottawa, Canada; Beijing, between Chicago and Washington; Xi'an, near St. Louis; Changsha, south of New Orleans; and Guangzhou, off the western tip of Cuba.[8]

The arable land of China constitutes about 15.3 percent of the total land area of 973 million hectares (2.3 billion acres). Cultivated crops are regularly grown on only 11 percent of the total land, or about 107 million hectares.[9] Multiple cropping practices in southern China provide the equivalent of about 150 million hectares for yearly crop production.[10] By comparison, about 21 percent of the land in the contiguous United States is arable and about 116 million hectares are regularly planted.[11] Land in the remainder of China is used for grazing (about 18 percent), grasslands (about 28 percent), and about 300 million hectares (roughly 30 percent) is marginal agricultural land or former forests that are now abandoned.

Dwight Perkins has provided estimates of the population growth and the extent of land cultivation in China over the last six hundred years.[12] As shown in figure 1, the rate of population growth closely paralleled the growth of cropland until about 1905. The relative positions of these

---

[7] H. Fullard, ed., *China in Maps* (London and Chicago, 1972).
[8] Ibid.
[9] Wortman, "Agriculture in China"; and FAO, *FAO Production Yearbook*, vol. 35, 1981.
[10] Wortman, "Agriculture in China"; and Hsu, *Food for One Billion*.
[11] FAO, *FAO Production Yearbook*, vol. 35, 1981.
[12] Perkins, *Agricultural Development in China*.

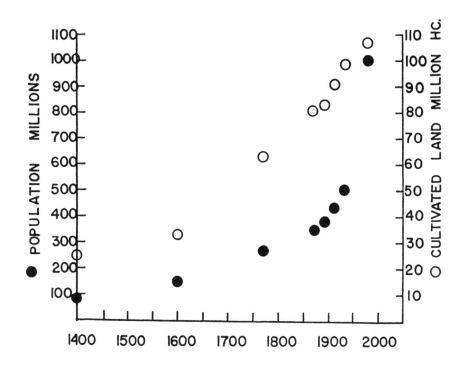

Figure 1. Comparison of Increases of Cultivated Land and of Total Population in China, 1400–1980.

Sources: R. C. Hsu, *Food for One Billion: China's Agriculture since 1949* (Boulder, Colo., 1980); D. H. Perkins, *Agricultural Development in China, 1368–1968* (Chicago, 1969); Food and Agricultural Organization (FAO), *FAO Production Yearbook,* vol. 35, 1981 (Rome, 1982).

two curves since 1949 demonstrate the dramatic improvement in the "carrying capacity" of Chinese agriculture since the birth of Communist China. By 1980, China was able to feed its billion people on 0.097 hectares of arable land per capita, a ratio that compares more than favorably with its neighbors: N. Korea, 0.12; S. Korea, 0.053; India, 0.24; Pakistan, 0.23; and Thailand, 0.34.[13]

Work Force

Agriculture is China's most important occupation by far. In 1981, 58.9 percent of a total population of 1,007,750,000 was involved in agriculture, with 27.3 percent, or 275,906,000, classified as economically active.[14] The United States, by comparison, in 1981 had 2.9 percent of its population (4,736,000) involved in agriculture, of which 0.94 percent (2,157,000) was judged economically active.[15] As in the rest of the world, the percentage of Chinese in the agricultural work force seems to be steadily decreasing with the advent of mechanization and the lure of the metropolis. In China the percentage of the population involved in agriculture in 1981 (58.9) was down nearly 9 percent from the 1970 level of 67.8.[16] In many respects the success of China's agriculture throughout history is a reflection of its enormous human resources, which have made possible the vast reconstruction of terraces and irrigation networks that allowed the arid regions of central China to become productive. The same abundance of eager fingers provided the skills for intensive cultivation and the multiple cropping system, as well as for a general system of nonpolluting pest control based on hand picking of weeds, diseased plants, and insect pests.[17]

Irrigation

The massive coordination and unification of water control that began in the loess plateau of China more than four thousand years ago has

---

[13] FAO, *FAO Production Yearbook*, vol. 35, 1981.
[14] Ibid.
[15] Ibid.
[16] Ibid.
[17] National Academy of Sciences, Committee on Scholarly Communication with the People's Republic of China (CSCPRC), "Insect Control in the People's Republic of China," Report No. 2 (Washington, D.C., 1977).

been a definitive factor in the success of Chinese agriculture; indeed, China has been described in the broadest terms as a "hydraulic civilization."[18] The recurrent theme of Chinese history has been the devastating cycle of flood and famine. The large-scale flood control efforts that were necessary for the settlement and agricultural development of the North China plain were made possible by the mobilization of labor on an unprecedented scale, as were the centuries of large-scale terracing and proliferation of irrigation so characteristic of the loess plateau and its adjacent river basins. Since Liberation the development of massive flood control projects has made large-scale destructive flooding a rarity.

By 1949, approximately 26.7 percent of the 100 million hectares of cultivated land was irrigated. The percentage of irrigated land has steadily increased, and by 1980 amounted to 47.3 percent of cultivated land.[19] This dramatic increase in irrigated land has contributed substantially to the success of multiple cropping, which has added almost one-third to the production capacity of Chinese agriculture and made China the world's largest producer of rice.[20]

Fertilization

Abundant fertilization has always been the secret of maximal agricultural production, whether it be the time-honored varieties of human and animal manure or the synthetic nitrates and ammonia of the "green revolution." Throughout history, the Chinese have practiced total recycling of nitrogen on a scale almost unimaginable elsewhere. The practice of nitrogen conservation is familiar to the incredulous Occidental traveler in China. Rural toilets are located directly over fishponds, and Chinese aquacultural yields are the highest in the world. In agricultural communes all human waste is carried to large underground tanks, where the heat of fermentation destroys pathogenic bacteria before the liquefied product is pumped into the fields. Animal manure is conserved as rigorously as human waste, even to the extent of employing collection bags hung under the tails of beasts of burden.[21]

---

[18] Buchanan et al., *China: The Land and the People*.
[19] Hsu, *Food for One Billion*.
[20] Wortman, "Agriculture in China."
[21] R. L. Metcalf, "China Unleashes Its Ducks," *Environment* 18, no. 9 (1976): 14–17.

The utilization of human and animal manure has increased dramatically since the years immediately following Liberation, more than doubling in a quarter-century (table 2).

Before 1960, soil fertility in Chinese agriculture was totally dependent upon these traditional uses of human and animal wastes, together with green manures, the planting of leguminous crops, and the spreading of nitrogen-rich river-bottom silt on fields. The advent of the "green revolution" in China, through the introduction of high-yield strains of dwarf rice heavily dependent upon nitrogen fertilizer for maximum yields, produced a major demand for synthetic fertilizers. By 1965, high-yield rice was grown on 3.3 million hectares, and by 1973, on 6.7 million hectares, or about 20 percent of all rice cropland.[22] By 1977 these high-yielding strains were being sown on 80 percent of the rice acreage.[23]

During the Great Leap Forward, massive efforts were made to produce ammonium bicarbonate fertilizer (17 percent nitrogen) in small factories. These factories produced 30 percent of the chemical fertilizer in 1965, 50 to 60 percent in 1972, and 45 percent in 1979, when there were 1,533 small nitrogen fertilizer plants.[24] By 1978 large fertilizer plants using natural gas and petroleum were producing 2.7 million metric tons of solid nitrogen fertilizer.

From 1961 to 1979, the production and import of chemical fertilizers multiplied by a factor of 21 (table 3).

With the increased availability of nitrogen fertilizers has come increased utilization: in 1961–65 applications averaged 13.3 kilograms per hectare; in 1972, 45.5; in 1977, 74.3; and in 1979, 109.2.[25] Utilization in 1979 was substantially greater than the world average of 77.1 kilograms per hectare, but still far short of the figure of 477.7 obtaining in Japan.

---

[22] Wortman, "Agriculture in China."
[23] Wortman, "Agriculture in China."
[24] FAO, *FAO Fertilizer Yearbooks*, 1978, 1976–79 (Rome, 1979, 1980).
[25] Wortman, "Agriculture in China"; Hsu, *Food for One Billion*; and J. R. Harlan, "Plant Breeding and Genetics," in *Science in Contemporary China*, ed. L. A. Orleans (Stanford, Calif., 1980), 295–312.

Table 2

Utilization of Human and Animal Manure in China's Agriculture
(in Million Metric Tons)

	Human	Hog	Draft Animal	Total
1952	186	130	422	738
1960	250	169	490	909
1970	346	379	609	1334
1977	398	492	763	1653

Source: Hsu, *Food for One Billion,* 54.

Table 3

Production and Importation of Chemical Fertilizers into China
(in Million Metric Tons)

	Domestic Production	Importation	Total
1961	0.36	0.23	0.59
1965	1.48	0.64	2.12
1971	3.05	1.37	4.42
1973	4.23	1.59	5.82
1977	7.24	1.48	8.72
1979	10.65	1.73	12.38

Source: Hsu, *Food for One Billion,* 56; FAO, *FAO Fertilizer Yearbooks, 1978, 1976-79* (Rome, 1979, 1980).

Plant Breeding

Improvement in crop varieties—the "green revolution"—is generally considered one of the important factors in the success of Chinese agriculture.[26] A too-narrow focus on the last quarter-century's successes, however, obscures the achievements of past millenia. The sheer magnitude of China's efforts in "hydraulic agriculture" and the millions of human beings involved throughout history suggest that Chinese efforts in crop domestication and improvement have long been among the most successful and extensive in the world. At least four stages in crop improvement are recognized: (a) domestication, (b) selection of optimal varieties, (c) importation of new crops from other continents, and (d) hybridization and other genetic manipulation.

As we have seen, domestication of major crop varieties in China occurred between 4000 and 3000 B.C. As irrigation practices developed and new land came under cultivation, population pressures demanded changes in crop development and growing time. Early-maturing varieties of rice were introduced from Indochina around A.D. 1100. These "Champa" strains matured in about 100 days, as compared to 180 days for the traditional strains, and were particularly useful in areas subject to seasonal flooding. Because they required less water, it was possible to extend rice culture to higher land, previously unsuitable for rice growing. The innovation of the Champa strains of rice made possible the double cropping of rice so characteristic of South China.[27]

Throughout succeeding centuries, the selection of even more rapidly maturing varieties produced wholly revolutionary changes in rice culture. By 1200, successful rice culture had been extended to terraced rice paddies in the hilly land of the lower Yangzi valley. The use of these new early-maturing varieties, said to have doubled the food output of Chinese agriculture over the past millenium, must have been a major factor in the quadrupling of the Chinese population during the Qing (Manchu) dynasty (A.D. 1644–1912).[28]

The introduction of maize, the potato, the sweet potato, and the peanut during the sixteenth and seventeenth centuries made it possible

---

[26] Buchanan et al., *China: The Land and the People*.
[27] Ibid.; and Perkins, *Agricultural Development in China*.
[28] Perkins, *Agricultural Development in China*.

to expand agriculture to upland areas of North China and the Yangzi valley, and these crops added substantially to the total food supply.[29]

The third major crop improvement occurred during the 1960s, following a period of stagnation. Finding it almost impossible to feed its burgeoning population, the Chinese developed their own short-statured, high-yielding "indica" varieties of rice, the first of which was released in Guangdong Province in 1960, six years before the famous IR-8 variety's release by the International Rice Research Institute (IRRI).[30] Acceptance was rapid, and the new, high-yielding varieties were being grown on 3.3 million hectares by 1965, 6.7 million by 1973, and about 27 million by 1979, or 80 percent of the rice acreage.[31] With adequate fertilizer and water, yields rose to five to six metric tons per hectare and required only 110 to 115 days to mature. (The impact of these new varieties on grain production can be seen clearly in figure 2.) This doubling of the cereal supply resulted in the dramatic increase in the "carrying capacity" reflected in China's population increase from six hundred million to one billion over the period from 1952 to 1982.

Since 1974 Chinese scientists have begun to use the world resources of germ plasm available from IRRI and other centers; further advances in the "green revolution" are undoubtedly underway. The breeding of crop varieties resistant to common plant diseases and insect pests has a high priority.[32]

During the Cultural Revolution, plant breeding programs were disrupted. Virtually no improvements occurred from 1966 until 1973, when a new hybrid rice was developed by crossing a wild rice from Hainan Island as the female parent. This rice increased yields by about 20 percent (750–1500 kilograms more per hectare) and had better drought resistance. It is less resistant to disease, however, and matures more slowly. In 1976 hybrid seeds were distributed for use on 130,000 hectares, an area that expanded to 2 million hectares in 1977 and to more than 5 million hectares in 1978. The increased yield from these plants

---

[29] Wortman, "Agriculture in China"; and Harlan, "Plant Breeding and Genetics."
[30] Hsu, *Food for One Billion*.
[31] Harlan, "Plant Breeding and Genetics."
[32] Hsu, *Food for One Billion*.

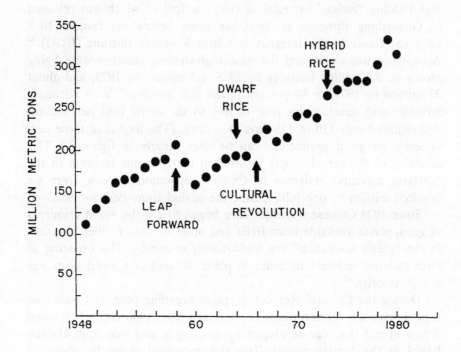

Figure 2. Increase in the Total Production of Grains in China, 1949–1980.

Sources: Hsu, *Food for One Billion;* FAO, *FAO Production Yearbook,* vol. 35, 1981.

accounted for more than one third of the total increase in rice production in 1979.[33]

Mechanization

Abundant human power has been one of the great historical strengths of Chinese agriculture. But in 1955, Mao Zedong declared that the mechanization of farming is the fundamental answer to increasing demands for food production. At that time Chinese leaders were influenced by the Soviet experience in collective farming and distracted by visions of "big farms and big tractors." In retrospect, there seems to have been a general failure to recognize that although mechanization can increase labor output, it cannot increase either the acreage of arable land or the inputs of fertilizer and better seeds that are the key ingredients of increased crop production.[34] The debates among Chinese leaders on the issue of mechanization contributed to an indecisive policy that became steadily more troubled during the confusion of the Great Leap Forward; agricultural mechanization soon lost its priority. During the Cultural Revolution, emphasis was finally shifted away from the production of large tractors and combines to the manufacture of smaller machinery, such as hand tractors and rice transplanters. As a result, the production of hand tractors increased almost sixfold, from 136,565 per year in 1969–71 to 740,000 in 1981.[35] These small tractors and rice transplanters were well suited to multiple cropping of rice combined with wheat, which is especially labor-intensive because of the overlap of harvesting and transplanting of the next crop.[36]

Goals for mechanization announced in 1978 aimed for 85 percent mechanization of agriculture, forestry, animal husbandry, and fisheries by 1985. There are, however, fundamental problems with agricultural mechanization, especially in countries with abundant manpower. Mechanization can lead to surplus rural labor of 50 percent or more. This is especially serious in China, where government restrictions prevent the movement of labor from rural to urban areas. The production increases

---

[33] Kang Chao, *Agricultural Production in Communist China* (Madison, Wis., 1970).
[34] FAO, *FAO Production Yearbook*, vol. 35, 1981.
[35] Hsu, *Food for One Billion*.
[36] Ibid.; and Chao, *Agricultural Production in Communist China*.

brought about by triple cropping do not necessarily increase total agricultural output, since lower average crop yields, greatly increased demand for fertilizer, and increased peak labor requirements are the less advantageous "side effects" of such intensive land use. In some areas of China total output and income declined, and there has since been a return to double cropping. The high capital cost of farm machinery and its maintenance and repair have prevented many communes from reaching a high level of mechanization.[37] These "farm problems" are not unfamiliar in the United States today.

In short, mechanization has not developed as rapidly as planned. The percentage of cropland plowed by tractors was only 33 in 1978 and 41 in 1981. In 1981, rice transplanters were used on only 0.7 percent of the paddies and harvesters on only 2.6 percent of the grain land.

Food Production

The Chinese Government announced in 1971 that China was now self-sufficient in food production; subsequent observations by independent observers have verified this statement. Today China feeds more than one billion persons, almost 25 percent of mankind, on the proceeds of 7.5 percent of the world's arable land. This has been described as "one of the monumental achievements of our times."[38]

The Chinese diet consists largely of cereals; between 1978 and 1980, it was estimated that approximately 90 percent of per capita intake of calories and 82 percent of the protein intake were of vegetable origin.[39] The emphasis on vegetarian nutrition is an important part of China's success in feeding its population, as a cereal diet consumed directly by human beings is roughly ten times more efficient than feeding cereals to hogs, sheep, or cattle for an animal protein diet. Moreover, dietary overemphasis on animal protein and fats is the cause of a number of human health problems, e.g., atherosclerosis, high blood pressure, and colon cancer, all of which are more common in the Western world than in China or other Asian nations.

---

[37] Buchanan et al., *China: The Land and the People.*
[38] FAO, *FAO Production Yearbook*, vol. 35, 1981.
[39] Wortman, "Agriculture in China"; CSCPRC, "Insect Control"; B. E. Ford, H. L. Bissonnette, J. G. Horsfall, R. L. Miller, D. Schiegel, B. G. Tweedy, and L. G. Weathers, "Plant Pathology in China, 1980," *Plant Diseases* 65, no. 9 (1981): 706–14.

A quantitative look at the total production of cereals in China since 1949 is shown in figure 2. It is evident that production has increased by 300 percent since 1949 and by 72 percent since 1971, when self-sufficiency in food production was announced. Despite the substantial population growth over this period (see fig. 1), China has gained in per capita intake of calories over the past twenty-five years, and the relative increase is appreciably greater than for most of its Oriental neighbors (see table 4). The present caloric intake is above the minimum considered essential.

The growth curve for cereal production (fig. 2) has interesting perturbations that demonstrate clearly the influence of both scientific developments (i.e., the introduction of early-maturing, dwarf strains of rice in 1964, and of hybrid rice in 1973) and of political decisions (e.g., the Great Leap Forward of 1958 and the Cultural Revolution in 1966). The fortunate combinations of manpower, irrigation, fertilization, and plant breeding that made possible China's overall success in food production have already been discussed.

Production statistics for foodstuffs other than cereals have also shown constant growth since 1949. The data presented in table 5 are further indications of the effective growth of Chinese agriculture and explain the repeated observations of Westerners since 1974 that the Chinese people appear to be well fed and well nourished.[40] Despite these impressive achievements in cereal production, however, there remains considerable opportunity for improvement. While the average Chinese yield of paddy rice in 1980, 3,717 kilograms per hectare, was considerably above the Asian average of 2,646 kilograms, it was substantially below that of South Korea (6,556), North Korea (6,150), Japan (6,240), and Taiwan (4,459).[41]

Plant Protection in China

The combined attack of pests, i.e., weeds, fungi, insects, birds, and rodents, on China's agricultural crops was estimated in 1937 to have caused a 10 to 20 percent loss of grain production.[42] Cramer[43] in 1967

---

[40] FAO, *FAO Monthly Book of Statistics, 1980* (Rome, 1980).
[41] J. L. Buck, *Land Utilization in China* (Nanying, 1937), 4.
[42] H. H. Cramer, "Plant Protection and World Crop Production," *Pflanzenschutz Nachrichten* 20: 1–524 (Leverkusen, W. Germany, Farbenfabriken Bauer, 1967).

Table 4

Per Capita Caloric Intake in China and Neighboring Countries

(in Calories per Capita per Day)

	1966–68	1975–77	1978–80	Relative Increase
China	2,112	2,370	2,472	1.17
India	1,860	1,893	1,998	1.07
Pakistan	2,015	2,226	2,300	1.14
Thailand	2,255	2,252	2,301	1.02
Japan	2,675	2,798	2,916	1.09

Source: FAO, *FAO Production Yearbook,* vol. 35, 1981, 248–49.

Table 5

Production of Foodstuffs in China
1969–1981

(in Thousand Metric Tons)

	1969–71	1979	1981	Relative Increase
Rice	109,853	146,959	146,087	1.33
Maize	32,376	60,149	61,601	1.90
Potatoes	12,362	15,536	15,039	1.22
Pulses	6,478	6,740	6,856	1.06

(in Thousands)

Pigs	188,734	305,607	310,251	1.64
Sheep	81,987	96,397	105,200	1.28
Chickens	634,348	798,360	861,393	1.36

Source: FAO, *FAO Production Yearbook,* vol. 35, 1981.

presented data based on comparable losses in Southeast Asia to suggest that the percentage losses to insects and plants diseases, respectively, in China were: rice, 15 and 8 percent; corn, 12.4 and 9.4 percent; wheat, 5 and 9 percent; cotton, 16.1 and 12 percent; sugar cane, 20.1 and 19.2 percent; and apples, 13.2 and 13.7 percent. These data are almost certainly exaggerated and make no allowance for the very superior system of plant protection in post-revolutionary China. However, in a country struggling to feed an immense population still growing at about 1.6 percent per year, protecting crops from pest attack is an obvious way to increase the available food supply without creating new agricultural land, improving irrigation, or increasing fertilizer production. Therefore, the Chinese have placed great emphasis on the science of plant protection.

It is estimated that more than 3.5 million persons are regularly engaged in the network of agro-scientific activities related to plant protection.[44] As shown in figure 3, this network extends from the Ministry of Agriculture, which receives input from the Agricultural Research Institute, to science and technology groups at the provincial, district, and county levels; then to the vast number of agricultural communes, with their science and technology groups; and finally to the production brigades and production teams. Plant protection specialists at this lowest level forecast pest outbreaks, sample pest populations to determine action thresholds for treatment, popularize new resistant crop varieties, and actually produce large quantities of agents for biological control, including microbial insecticides, such as *Beauveria bassiana* and *Bacillus thuringiensis*, beneficial parasites, such as *Trichogramma spp.*, and predators, such as *Coccinella septempunctata*.[45]

Coordination by the Ministry of Agriculture appears to be limited largely to the evaluation of new pesticides and the initiation of training programs. The forecasting of pest outbreaks has been intensified through the provincial academies of agriculture by means of their close co-

---

[43] B. Berner, "The Organization and Economy of Pest Control in China," Research Policy Studies, Discussion Paper No. 128 (Lund University, Sweden, 1979).

[44] CSCPRC, "Insect Control"; and R. L. Metcalf and A. M. Kelman, "Plant Protection," in *Science in Contemporary China*, ed. L.A. Orleans (Stanford, Calif., 1980), 313–44.

[45] CSCPRC, "Insect Control."

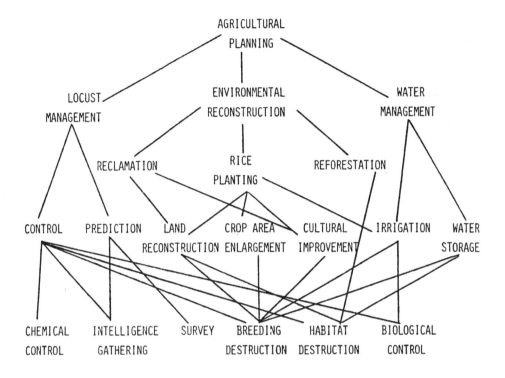

Figure 3. Components of Chinese Program for Pest Management of the Migratory Locust *(Locusta migratoria manilensis)*.

Source: National Academy of Sciences, Committee on Scholarly Communication with the People's Republic of China (CSCPRC), "Insect Control in the People's Republic of China," Report no. 2 (Washington, D.C., 1977).

ordination with brigades and communes. The accumulated information is applied to both short- and long-term forecasts, and in many provinces there is daily radio communication between the provincial forecasting center and hundreds of brigades. In 1975, it was stated that there were 186 forecasting stations in Jilin Province.[46] Emphasis is placed on life histories, population fluctuations, and migrations of important pests; their key natural enemies; quantitative sampling methods; and development of economic thresholds.[47] It is no exaggeration to state that virtually every peasant working in agriculture has some involvement in observing, reporting, and destroying pests. This input from on-site personnel, moving upward through the plant protection specialists of the agro-scientific network, has given China an efficient, innovative, and flexible plant protection system that is the envy of the rest of the world. The capabilities of this system are illustrated by the two case histories summarized below.

Control of the Migratory Locust in China, 1950–1975

The migratory *Locusta migratoria manilensis* has been the most destructive pest of Chinese agriculture throughout history. Recorded locust plagues date back to 707 B.C., and more than eight hundred epidemic outbreaks have occurred, chiefly in the flood plains of the Huang, Huai, and Yangzi rivers and in the coastal regions of South China.[48]

When breeding gounds become crowded, these locusts migrate in vast swarms of hundreds of millions of individuals that may weigh up to five hundred tons aggregate and consume their weight in vegetation every day. Successive outbreaks of locusts have occurred every three to fifteen years in China and have lasted from twelve to thirty months. In 1929, a massive locust outbreak devastated more than 4.5 million hectares of cropland and caused illness, starvation, and death as well as wholesale migration of millions of people away from the affected area.[49]

Efforts at locust control date back to the Shang dynasty (1523–1027 B.C.). A later poem describes the use of flame to destroy locusts. In the Han (202 B.C.–A.D. 220), ditches were used to trap immature locusts.

---

[46] Ibid.; and Metcalf and Kelman, "Plant Protection."
[47] Metcalf and Kelman, "Plant Protection."
[48] Ibid.
[49] Ibid.

During the Tang dynasty (A.D. 618–906), burning and trapping were combined, and during the Song (A.D. 960–1279), soil was dug up to destroy locust eggs. By 1182 a law required citizens to collect and destroy locusts, very likely the first legal requirement for a pest control activity.[50]

The biology and ecology of locust behavior was well understood in the Ming dynasty (A.D. 1360–1644), when texts were published showing the insect's habit of migrating upwind and early in the morning, its reluctance to fly at dusk or in the afternoon after mating, and the tendency to fly low after rain and toward flames. By about 1600, a major step forward was indicated by Shen Guangjing's description of the high prevalence of locust invasions in Hebei, Shandong, and Henan provinces and his call for concerted action at a national level to control the pest.[51]

After Liberation, a massive national effort to control migratory locusts was begun by Chinese entomologists under the direction of the Academy of Sciences and the Ministry of Agriculture. Locusts infest an area of more than one million hectares, where they breed in reeds growing on the flood plains of rivers and lakes. Thus, locust outbreaks are correlated with periodic flooding and droughts. Intensive ecological study identified the primary breeding centers and secondary centers from which dispersion occurs.[52] During a major outbreak in 1951, millions of persons were mobilized to dig up and destroy locust eggs, and to surround, swat, and burn the immature locusts. Extensive aerial applications of benzene hexachloride (BHC) insecticide were made from 1951 to 1963.[53] In the long run, these strategies were deemed impractical and supplanted by more ecologically oriented efforts, directed by more than sixteen monitoring and forecasting stations established in the North China plain. From 1958 on, enormous efforts were directed at converting locust breeding areas into rice paddies, forest belts, and pastures. More than one million hectares of hilly farmland in the Huang River valley were converted into terraced fields, four million check dams

---

[50] Ibid.

[51] Chin Chun-teh, "Combining Ecological Reform with Control in the Elimination of Migratory Locust Outbreaks in China" (personal communication, 1981).

[52] Metcalf and Kelman, "Plant Protection."

[53] Ibid.

were built, and trees and grasses were sown on four million hectares. In the Weishan Lake area (Shandong Province), the landscape was transformed with forty million man-days of labor and 550 million cubic meters of concrete to build check dams, irrigation systems, and ditches.[54]

More than 120 million persons participated in such massive efforts for locust control, and the periodic outbreaks have now been virtually eliminated. This effort is undoubtedly the greatest pest control program of history. Its success was due to a scientifically directed, imaginative, and practical approach for the environmental control of locusts that utilized the techniques (fig. 3) of biology, ecology, meteorology, and water management, together with established cultural practices and heavy reliance on forecasting.[55]

Rice Pest Management in Henan Province in the 1980s

Rice is traditionally vulnerable to a large variety of insect pests; at least eighty-nine important ones were discussed with the U.S. Insect Control Committee in 1975.[56] Throughout the centuries, the major pests have been stem borers—the rice stem borer (*Chilo suppressalis*), the paddy borer (*Tryporyza incertulas*), and the purple stem borer (*Sesamia inferens*)—as well as the oriental armyworm (*Mythimna separata*), the lawn armyworm (*Spodoptera mauritia*), and the rice green stink bug (*Nezara viridula*).

After 1949, vast quantities of DDT and BHC insecticide (more than 130 million pounds of BHC in 1958) alone were produced in China, and as much as 70 percent was more or less haphazardly applied by aircraft.[57] These broad-spectrum insecticides decimated the natural enemies of the leafhoppers and planthoppers, especially *Lycosa* spiders and such hymenopterous parasites as *Trichogramma*. The ecology of the rice paddy was still further altered by the introduction of the dwarf *indica* varieties of rice. These strains lacked the insect pest-resistant genes developed over the millenia by the traditional *japonica* varieties. The

---

[54] CSCPRC, "Insect Control"; and Metcalf and Kelman, "Plant Protection."

[55] CSCPRC, "Insect Control."

[56] Cheng Tsien-hsi, "Insect Control in Mainland China," *Science* 140 (1963): 269–77.

[57] Wortman, "Agriculture in China"; Hsu, *Food for One Billion*; and Harlan, "Plant Breeding and Genetics."

dwarf *indica* required greatly enhanced applications of synthetic nitrogen fertilizer as well as regular applications of insecticides.[58]

These changes in the ecology of rice paddies have had calamitous effects on rice culture in the Orient, both through development of insecticide resistance in many rice pests and by destruction of their natural parasites and predators. Pests formerly considerd only slightly destructive, including the green rice leafhopper (*Nephotettix cincticeps*), the brown planthopper (*Niloparvata lugens*), and the white-backed planthopper (*Sogata furcifera*), have become dominant secondary pests, especially destructive because they transmit a complex of virus diseases of rice, e.g., yellow dwarf, yellow stunt, and common dwarf. These secondary pests developed resistance to DDT and BHC, and the organophosphorus insecticides parathion and methyl parathion were introduced for control. Increasing numbers of applications were made with damaging effects on worker health and fish culture.[59] This experience was typical throughout Southeast Asia; in some areas, such as Japan, the use of insecticides increased 30-fold with rice yields increasing only 1.5-fold, despite the new varieties of rice and heavy fertilization. As a consequence of this "pesticide treadmill," the rice pests developed multiple resistance to organochlorine, organophosphorus, and carbamate insecticides. The green rice leafhopper, for example, is resistant to at least eighteen and the brown planthopper to at least fourteen.[60]

The simplistic solution was to spray more frequently. From four to seven insecticide applications were made during the 1970s, as compared to one to two in the 1960s. Thus, the important insecticide resources to control these pests are now virtually depleted, and this is a major factor in the catastrophic situation that now exists for rice insect pest control in Southeast Asia.[61] The brown planthopper is almost uncontrollable,

---

[58] R. L. Metcalf, "Changing Role of Pesticides in Crop Protection," *Annual Review of Entomology* 25 (1980): 219–56; and K. Kiritani, "Pest Management in Rice," *Annual Review of Entomology* 24 (1979): 279–312.

[59] R. L. Metcalf, "Trends in the Use of Chemical Insecticides," Proc. FAO/IRRI Workshop on Judicious and Efficient Use of Insecticides on Rice. International Rice Research Institute (Los Banos, Laguna, Philippines, 1984).

[60] Kiritani, "Pest Management in Rice."

[61] Metcalf, "Changing Role of Pesticides."

and five biotypes are now known, resistant not only to insecticides but also to successive new introductions of host-resistant dwarf rices.

The resiliency of the Chinese system of integrated pest management is shown by the system developed in China's rice bowl, Henan Province, to cope with this situation. Henan grows about two million hectares (roughly 6 percent) of China's rice; in some areas, such as Xiangying County, rice comprises more than 80 percent of the agricultural crops. Henan's rice insect pests, in order of importance, are the brown planthopper, the rice leaf roller (*Cnaphlocrocis medinalis*), and the rice stem borer. The brown planthopper first became a problem in about 1970, when a new strain of rice was introduced from the Philippines. Dusting with 3 percent BHC and 3 percent methyl parathion was widely applied to the double-cropped rice, with applications to the first crop at the tillering stage in May and the heading stage in June, and to the second crop about the middle of August, and twice in September. The costs were excessive and the results uncertain.[62]

The development of a successful pest management scheme was preceded by intensive study of rice paddy ecology. It was discovered that there are as many as twenty species of spiders found in untreated rice paddies that are the key predators of the brown planthopper and other pests. Strangely enough, a Chinese proverb at least 2000 years old emphasized that spiders were the rice growers' friends—lore which had been all but forgotten in the rush to develop a modern, chemically oriented rice culture.[63] The new pest management scheme, which is producing very acceptable yields of rice (5,250 kilograms per hectare for the first crop and 4,500 for the second), has reduced the average insecticide applications from five to two, one to each crop toward the end of tilling. The cost of the insecticide applications has been reduced from five to seven *yuan* per *mu* to less than one.[64]

This very successful rice pest management program incorporates the following components: a reduction in nitrogen fertilizer from 30 to 10 kilograms per *mu*; a change in insecticides to a 3 percent mixture of carbofuran (a systemic) and methyl parathion and a reduction in applications; a reevaluation and raising of the economic thresholds for

---

[62] Li Shaoshi, "Henan Pests Forecasting Station, Changsha, Henan, China" (personal communication, 1984).

[63] Ibid.

[64] Ibid.

insecticide application by two- to three-fold; development of hybrid resistant varieties of rice; and reduction in weeding of the rice paddies to provide refugia for the beneficial spiders. The key to this new program is clearly the preservation of the beneficial spider population by changing the insecticide; by treating only when necessary, and then with localized spot treatments of infested portions of the paddy; and by providing refugia for the spiders.

The overall pest management program is effective also because of the detailed organization of the plant protection service and the abundant farm labor force. The program was developed by professional biologists from Henan Teacher's College and the Henan Province Plant Protection Station. National and provincial pest forecasting and weather agencies provide technical information, using on-line radio to communicate directly with plant protection specialists and farmers. The lower echelons of this program are tightly interconnected: Xiangjin County has fifteen technical specialists, and each village or town has a plant protection service group of two or three technicians to service an average of one to two thousand hectares. Each farmer's control working group has one trained specialist to sample and monitor rice pest populations in relation to the economic threshold and to provide data for the provincial forecasting.[65]

This is an outstanding example of Chinese organization in response to a serious threat to the food supply. The integration of pest management to preserve high yields of rice while decreasing insecticide applications stands in marked contrast to many other countries in the Orient, where the response to the increased threat from rice pests has generally been more frequent spraying with a larger and larger array of pesticides. The end result has been an ever-widening circle of multiple resistance in the insect pests and increasing development of serious secondary pests.[66]

Research and Education in Agriculture

China's effort in agricultural research has been enormous. Research and development in agriculture and natural resources received 24 per-

---

[65] Ibid.
[66] Metcalf, "Trends in the Use of Chemical Insecticides."

cent of the total national expenditure in 1973–78 and involved 45 percent of the total research manpower.[67] In the whole of China, more than fourteen million farmers are estimated to participate in agro-scientific experimental programs, monitoring and preventing plant diseases and insect attacks, and popularizing the new technology.[68] Agriculture is a prime target for application of the Four Modernizations that sought by 1985 to: approach or reach world-class in science and technology; increase the number of professional scientific researchers to 800,000; build new centers for scientific experimentation; and complete a nationwide system of scientific technological research.[69] Agriculture's specific goal for implementation is the Eight-Point Charter, which calls for raising the level of scientific farming in regard to soils, fertilizer, water conservation, seeds, close-planting, plant protection, field management, and improved farm tools.[70]

The organization of agricultural research is complex and involves a network of institutions reaching from provinces and communes to production brigades and teams. This system has been aptly described as "specialization and coordination in a rationally distributed, countryside network for agricultural experiment."[71] A brief outline of this apparatus will illustrate its complexity and effectiveness and indicate the difficulty in understanding its ramifications and parameters, especially for the Western world.

Chinese Academy of Sciences (CAS)

The Academy of Sciences is a cabinet-level, premier research organization under the State Scientific and Technological Commission. Its functional groups are more than 120 national-level research institutes that have purview over the entire scientific domain.[72] At least 21 of these institutes are directly involved in basic research applicable to agriculture. Those institutes especially well known to the Western world include the Botany institutes in Beijing and Guangzhou, the Genetics and Microbiological institutes in Beijing, the Entomology and Plant

---

[67] Berner, "Organization and Economy of Pest Control in China."
[68] Ibid.
[69] Orleans, ed., *Science in Contemporary China*.
[70] Berner, "Organization and Economy of Pest Control in China."
[71] Ibid.
[72] Orleans, ed., *Science in Contemporary China*.

Physiology institutes in Shanghai, and the Hydrobiology Institute in Wuhan. Their publications, mostly appearing in the respective *Acta Sinica* divisions (botanica, entomologica, genetica, hydrobiologica, microbiologica, phytopathologica, phytophysiologica, and zoologica), represent the Chinese medium of basic exchange for scientific information with the rest of the world.

The Academy of Agricultural Sciences (CAAS)

This is the highest-level research organization under the Ministry of Agriculture, established in 1957. Its mission was defined as applied research relating to national problems of agriculture. By 1980, there were at least thirty-one institutes under the Academy of Agricultural Sciences, with titles encompassing agricultural economics, agricultural mechanization, animal husbandry, apiculture, biological control, fiber, citrus, cotton, irrigation, fruit trees, germ plasm, grasslands, oil crops, plant protection, sericulture, soil and fertilizer, sugar beets, tea, tobacco, vegetables, and veterinary sciences.[73]

Provincial Academies of Agricultural Sciences

At least seventeen of China's twenty-nine provinces and two municipalities have provincial academies responsible for research at the provincial level and for coordination of research and extension activities, such as plant variety evaluation and pest monitoring and forecasting. These provincial academies oversee and coordinate the work of a variety of comprehensive and special institutes.[74]

The Shanghai Academy of Agricultural Sciences, established in 1960, had by 1975 a staff of about 1,300, of whom 400 were classified as professional scientific researchers. The academy has research institutes in plant breeding, plant protection, horticulture, livestock, and soil fertilization.[75] The Jilin Academy of Agricultural and Forestry Sciences is responsible for eleven provincial research institutes, such as the well-known Institute for Plant Protection at Kungchuling, as well as seven comprehensive district institutes working in animal husbandry, agricultural machinery, crop breeding, crop cultivation, and plant protection.[76]

---

[73] Ibid.
[74] Berner, "Organization and Economy of Pest Control in China."
[75] CSCPRC, "Insect Control."
[76] Berner, "Organization and Economy of Pest Control in China."

Four-Level Network

The Cultural Revolution severely disrupted the basic and applied research endeavor. Professional societies were disbanded, journals ceased to be published, the Scientific and Technological Commission (STC) was abolished, and many research institutes formerly under central governmental control were either closed or decentralized. Revolutionary committees took over the administration of research institutes, and the conventional distinctions between research, education, and production were almost obliterated. Many scientists were forced to leave their laboratories and spend large blocks of time "learning from the people." The evil effects of this period, not only on scientific progress in general but on individual scientific careers, are well known and can readily be imagined.

Reason slowly reasserted itself, and by 1977 the Four Modernizations became the cornerstone of governmental policy, with modernization of science and technology as the key.[77] The STC was reestablished, and the CAS institutes were restored to their former concentration on basic research and large-scale framework. The "Four-Level Agroscientific Network" was established to adapt and communicate research results to the county, commune, brigade, and production team levels.[78] Through this network, advanced farming technology was to be administered by the masses and supported by the state, which has the responsibility for adapting new technology to local conditions, carrying out the necessary experimentation, and taking the mass line. The present situation is diagrammed in figure 4.

The majority of counties, communes, brigades, and teams have some type of research or extension unit. In 1979 Henan Province was said to have 1.5 million agrotechnicians, about 4 percent of the total agricultural population. In Jilin Province more than 200,000 persons were similarly involved, about 1 percent of the agricultural population.[79] In Jilin all counties had research institutes, staffed with twenty to fifty professionals trained at universities or technical schools. All of the communes in Jilin, some 1,100, had an agro-scientific station staffed with four or five technicians, and nearly 90 percent of the production brigades had experi-

---

[77] Orleans, ed., *Science in Contemporary China*.
[78] Berner, "Organization and Economy of Pest Control in China."
[79] Ibid.

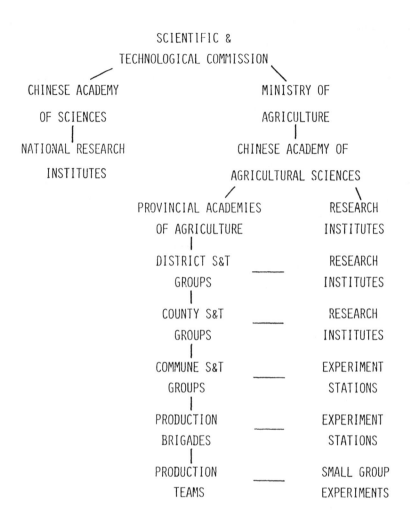

Figure 4. Scientific Research and Technology in Agriculture in China.

Sources: CSCPRC, "Insect Control"; B. Berner, "The Organization and Economy of Pest Control in China," Research Policy Studies, Discussion Paper no. 128 (Lund University, Sweden, 1979); R. L. Metcalf and A. M. Kelman, "Plant Protection," in L. A. Orleans, ed., *Science in Contemporary China* (Stanford, Calif., 1980), 313–44.

ment stations staffed with from two to fifteen technicians embodying the "three-in-one" principle of practical expertise in farming combined with technical skills. More than three-quarters of the production teams in Jilin had science and technology groups with as many as five members who participated in the normal farm work of the team but had specific extension responsibilities.[80]

Universities and Colleges

China's institutions of higher education, in contrast to the land-grant college system of the United States, never played a major role in agricultural research. Under the Four Modernizations, however, in 1978 approximately ninety key comprehensive universities and polytechnic colleges were selected for development as elite institutions to be given the best equipment and facilities, the most competent faculties, and the most academically qualified students.[81] These are presumably the academic institutions selected to train the scientific and technical manpower for China's modernization program and to develop scholarly contacts and exchange programs with academic institutions in the West. At present more than 20,000 Chinese students are enrolled in U.S. universities.

Zhongshan University, in Guangzhou, is an example of a major comprehensive university, begun by Sun-Yat-sen in 1924 and incorporating the campus of the former Lingnan University. It has departments of history, philosophy, economics, foreign languages, mathematics, physics, chemistry, metallurgy, geography, and biology (which includes biochemistry, entomology, genetics, microbiology, plant physiology, and zoology). The Biology Department, through its Institute of Ecological Entomology, is heavily involved in basic and applied research dealing with insect pests and, with professional staff trained in the United States and the Soviet Union, has developed, in cooperation with the Guangzhou Institute of Entomology, some of the most imaginative programs in biological control of rice insects to be found anywhere in China.[82] These sorts of interrelationships between university and government research institutes probably represent the future in China.

---

[80] Ibid.
[81] Orleans, ed., *Science in Contemporary China*.
[82] CSCPRC, "Insect Control."

Food Policy and Nutritional Status in China, 1949–1982[1]

Dean T. Jamison and Alan Piazza

Chinese development efforts from 1949 to about 1979 emphasized two main objectives: development of a heavy industrial base and elimination of the worst aspects of poverty.[2] The population of China in 1949 suffered a crippling burden of disease and premature death; perhaps the most striking success of China's subsequent antipoverty struggle has been to increase dramatically the level of life expectancy with a concomitant reduction in the burden of illness in the society. Public health measures, combined with reductions in malnutrition, improved water supplies, and close attention to hygiene and sanitation, have contributed to the increase in life expectancy from thirty-two to sixty-nine years in the period from 1950 to 1982; this level is only about six years lower than that

---

[1] This paper is based in part on previous work of the authors: D. T. Jamison and F. Trowbridge, "The Nutritional Status of Children in China: A Review of the Anthropometric Evidence," Technical Note GEN 17, Population, Health and Nutrition Department, The World Bank (Washington, D.C., 1984); and A. Piazza, *Food Consumption and Nutritional Status in the People's Republic of China* (Boulder, Colo., 1986).

[2] For an overview of the Chinese economic system and its recent performance, see the World Bank report entitled *China: Socialist Economic Development* (Washington, D.C., 1983).

found in the industrialized market economies.[3] Growth in agricultural production and improvements in food distribution have substantially alleviated hunger as an important facet of poverty reduction as well as contributed to improved life expectancy. Nonetheless, it is important to bear in mind from the outset that progress in improving health and nutrition conditions has been far from uniform and that major rural-urban differences and differences among rural areas still exist.[4]

Although China's overall living standards have improved substantially during the period since 1949—indeed, they have improved somewhat more rapidly than have living standards in developing countries generally—it is important to emphasize that general improvements in living standards alone can account for only a fraction of China's achievements in nutrition and health. Table 1 presents selected development indicators for China and a number of other countries and groups of countries. Columns 1 and 5 indicate that, although per capita GNP is low in China, life expectancy is nonetheless high by comparison even with countries having substantially higher income levels. Income *growth* in China has been moderately rapid, but columns 1, 2, and 6 of table 1 suggest that neither the level nor growth rate of income explains China's improvement in life expectancy: the twenty-seven-year increase in life expectancy between 1960 and 1980 exceeds that of other countries sufficiently to indicate the importance of other factors. Potential other factors, beyond the activities of public health agencies, are education levels (column 8 of table 1), reductions in population growth rate (columns 3 and 4), availability of food (column 7), distribution of available food, and improvements in water supply and sanitation. China's achievements in each of these areas, relative to its income level, have undoubtedly complemented the efforts of public

---

[3] The figures used in this paper for infant mortality rates, population totals, life expectancy, total fertility rates, and other demographic indicators result from a demographic analysis prepared by the World Bank to model officially available data. See Kenneth Hill, "Demographic Trends in China, 1953–1982," Technical Note 85-4, Population, Health and Nutrition Department, The World Bank (Washington, D.C., July 1985).

[4] For a detailed discussion of health conditions and health service in China, see Dean T. Jamison, John R. Evans, Timothy King, Ian Porter, Nicholas Prescott, and Andre Prost, *China: The Health Sector* (Washington, D.C., 1984).

health authorities in effecting the mortality and morbidity reductions of the past thirty years.

Our purpose in this paper is to examine more carefully the role of food policy and increased food availability in improving health and alleviating the worst aspects of poverty in China. The paper begins with a review of food policy and food availability in China. It then turns to an assessment of secular trends and regional and other differences in nutritional status in China. The concluding section discusses issues for the future.

Food Policy and Food Availability

Food Policy

The Chinese Government moved rapidly during the 1950s to establish control over agricultural production and both domestic and international agricultural trade. Control over agricultural production and trade was secured through indirect (i.e., price policy and advance purchase contracts) and direct (quotas for sown area and crop production) agricultural planning and through the development of a state monopoly of agricultural marketing.[5] State control of agriculture was seen as a central feature of socialism and, more pragmatically, was intended as a means for accelerating the generation of foreign exchange from agricultural exports, for supplying raw materials to light industry, and for securing adequate food supplies for the growing urban population. Having established effective control of food production and distribution, the government food policy consisted of: (1) implementing an urban food rationing and subsidy system and rural grain transfer programs; (2) controlling the mix of agricultural production and food consumption; and (3) moderating local and regional food shortages through international and domestic agricultural trade.

---

[5] Agricultural planning and the state marketing system are discussed in Dwight Perkins, *Market Control and Planning in Communist China* (Cambridge, Mass., 1966); Nicholas Lardy, *Agriculture in China's Modern Economic Development* (Cambridge, England, 1983); and Francis Tuan and Frederick Crook, *Planning and Statistical Systems in China's Agriculture* (Washington, D.C., 1983).

Table 1

Selected Development Indicators, China and Other Countries

	Per Capita GNP, 1980 (1980 U.S.$) (1)	Growth Rate of Per Capita GNP, 1960–80 (% p.a.) (2)	Population Growth Rate 1960–80 (% p.a.) (3)
Low-income economies (excluding India and China)	230	1.0	2.5
India	240	1.4	2.2
Sri Lanka	270	2.4	2.0
China	290	3.6[a]	1.8
Pakistan	300	2.8	3.0
Indonesia	430	4.0	2.2
Thailand	670	4.7	2.7
Middle-income economies	1,400	3.8	2.4
Hong Kong	4,240	6.8	2.5
Nonmarket industrial economies	4,640	4.2	0.9
Industrial market economies	10,320	3.6	0.9

Sources: For countries other than China and for country groupings, see World Bank, *World Development Report 1982* (hereafter, *WDR82*) (Washington, D.C., 1982). For China, *WDR82* and *China: The Health Sector* (Washington, D.C., 1984). For daily per capita energy supply for countries other than China, see Food and Agriculture Organization, *Food Balance Sheets: 1979–81 Average* (Rome, 1984). For China, see A. Piazza, *Food Consumption and Nutritional Status in the People's Republic of China* (Boulder, Colo., 1986). *WDR82* defines "low-income economies" as those having a per capita income of $410 or less in 1980; 33 such economies are included in *WDR82* tables. The "middle-income economies" are those of developing countries that have per capita incomes between $410 and $4,510; this group includes 62 countries. The nonmarket industrial economies have incomes ranging from $3,900 to $7,180, and the industrial market economies have incomes ranging from $4,880 to $16,440.

Total Fertility Rate, 1980 (4)	Life Expectancy (Years) 1980 (5)	Gain between 1960 and 1980 (6)	Daily per Capita Energy Supply, 1979-81 Average (kcal) (7)	Adult Literacy Rate, 1977 (%) (8)
6.1	57	15	2,066[c]	34
4.9	52	9	2,056	36
3.6	66	4	2,251	85
2.5	**67**	**27[b]**	**2,525**	**66**
6.1	50	7	2,180	24
4.5	53	12	2,372	62
4.0	63	11	2,330	84
4.8	60	9	2,593[c]	65
2.2	74	7	2,771	90
2.3	71	3	3,382[c]	100
1.9	74	4	3,415[c]	99

[a] This figure for China is the growth rate of gross *domestic* product (GDP) minus the population growth rate. For the period 1960-80, GNP and GDP in China grew at approximately the same rate.

[b] 1960 and the adjacent years were periods of acute famine and turmoil in China, which resulted in substantially elevated mortality rates. The 27 year gain in life expectancy reported here is, therefore, based on an imputed 1960 life expectancy that is the average of the 1957 and 1963 life expectancies.

[c] The figure is the average for 1980-81.

Rationing. The urban rationing system has been in operation since November 1953. Under the rationing system most households receive ration coupons for grain and certain other basic foodstuffs; approximately 16 percent of China's population benefits from this system.[6] The rations vary by age, sex, occupation, and locality and are heavily subsidized. The rationing system insures that all entitled urban residents receive adequate supplies of food and has helped to reduce greatly urban malnutrition in China, an important accomplishment in marked contrast to the severe malnutrition of the urban poor in many developing countries. The rationing system has also been extraordinarily costly, however, absorbing about 4.2 percent of China's 1981 GNP. The urban food subsidy is estimated to have been worth 96 *yuan* per urban dweller in that year, or almost one-and-one-third times the total expenditure on health care.

There is no formal rationing system in rural areas, but grain is transferred from surplus to deficit areas as part of the government's effort to maintain a floor level of grain consumption. The floor level of grain consumption varies from province to province and, within provinces, from county to county. Typically, the guaranteed minimum is 200 kilograms of unprocessed grain per annum in rice-growing areas, and 150 kilograms in other areas.[7] Provinces unable to produce enough grain to maintain this floor level of consumption are allowed to import grain from other provinces. Grain-poor provinces (excluding China's three municipalities) collectively imported an average of between 4.5 and 5.5 million tons of grain per annum in the 1950s, and about 8 million tons per annum in the late 1970s. Inner Mongolia, Gansu, Shaanxi, Ningxia, Qinghai, Guizhou, Yunnan, Guangxi, and Guangdong together imported

---

[6] In 1980 there were an estimated 9.3 million nonagricultural workers officially classified as contract or temporary workers who were not entitled to rationed food. For further information on the implications of the rationing system, see Nicholas Lardy, "Agricultural Prices in China," World Bank Staff Working Paper No. 606 (Washington, D.C., 1983).

[7] A rice ration of 200 kilograms (unmilled) per annum would provide 1,400 kcal per day of energy and 25 grams of protein—approximately 64 percent and 69 percent, respectively, of the 1979 level of daily requirements in China. A wheat ration of 150 kilograms (unmilled) per annum would provide 1,250 kcal per day and 35 grams of protein, 57 percent and 97 percent of requirements.

more than 4 million tons of grain in 1979, greatly alleviating the shortfall of grain production in those provinces.[8] Although rural grain transfers have certainly greatly increased the proportion of the population whose basic food needs are met, it is important to emphasize that the system does not satisfy everyone's food needs. This was most notably the case during the food crisis of 1959–62, but, as will be discussed, undernutrition is still prevalent in many rural areas, particularly in western China.

Mix of Food Consumption. The government has also used its control of agriculture to manipulate effectively the mix of food production and consumption. Until quite recently, government policy has favored the direct consumption of grain over the consumption of such preferred foods as pulses, vegetable oils, and animal products. As a result, the average Chinese now consumes more food energy through the direct consumption of grain than do the inhabitants of almost any other country—only Koreans and Egyptians come close. On the other hand, China's per capita consumption of pulses, separated oils and fats, and animal products is quite low by international standards. It should be noted, however, that consumption of preferred foods has been increasing rapidly in recent years and the government plans to continue this increase for the foreseeable future.

Since grains produce the greatest quantity of food energy and protein per hectare of cropland, it is possible to increase gross nutrient availability by restricting the production of preferred foods.[9] Largely in order to maximize gross nutrient availability from the limited arable land available, then, China's agricultural planners have chosen to restrict the production of preferred foods and to emphasize the consumption of grain. The adoption of this strategy explains the unusual fact that China has been able to attain the level of the average middle-income country

---

[8] See Kenneth Walker, *Food Grain Procurement and Consumption in China* (Cambridge, England, 1984), 77–78, 184–92.

[9] At least four to five kilograms of grain are required to produce one kilogram of meat. Consequently, since meat has no more energy and only about 30 percent more protein than grain (on a per kilogram basis), meat production involves large losses of available energy and protein supplies. The major exception to this rule is meat from pasture-fed animals, but such animals do not figure prominently in China's livestock sector.

in terms of per capita availability of food energy and protein (but not fat), but that the mix of food consumption is still very much typical of the average *low*-income country. (Daily per capita nutrient availability in China and selected other countries is presented in table 1.)

Maximizing nutrient availability in this way may have reduced malnutrition in China in two ways. First, of course, increased nutrient availability means there is more food energy and protein for everyone. Second, restricting the consumption of preferred foods may have brought about a more equal distribution of nutrient consumption. This is so since, as the income elasticity of demand for preferred foods is by definition greater than that of less preferred foods, the distribution of consumption of preferred foods is likely to be less equal across income groups relative to the distribution for less preferred foods.[10] In effect, restricting the consumption of preferred foods should have had the effect of increasing the availability of affordable food for China's poor. It is important to note, however, that restricting the production of preferred foods may have also had nutritionally deleterious effects. Specifically, the chronic short supply of nutrient-dense weaning foods, which are typically derived from preferred foods, may have adversely affected the health of infants. Additionally, the steady drop in per capita availability of soybeans and other pulses since the mid-1950s has circumscribed the possibility of complementing proteins in the average diet. Though the trend was dramatically reversed beginning in 1979, per capita production of oilseeds also declined steadily after the mid-1950s and may have resulted in possibly unhealthy low levels of fat intake during the 1960s and 1970s.

International Agricultural Trade. A primary objective of China's international agricultural trade has been to generate foreign exchange in order to pay for imports of capital goods. China has maintained a positive agricultural balance of trade since 1949 in all years except 1961–64 and 1980–81, and planners have never considered food imports to be a viable long-run solution to domestic shortfalls in food production. The relatively large annual grain imports of more than ten million tons of recent years, which have been misunderstood as a sign

---

[10] The Food and Agriculture Organization (FAO) reports empirical estimates of the elasticity of demand for agricultural products from a number of countries in *Income Elasticities of Demand for Agricultural Products* (Rome, 1976).

of insufficient domestic grain production, are more likely explained by internal transport constraints and grain procurement shortfalls.[11]

Maintaining a positive agricultural trade balance, however, cannot explain the astounding fact that China increased net grain exports during the first two years of the food crisis of 1959–62. China's grain trade and its impact on domestic grain supplies are summarized in table 2. Annual net grain exports, which had averaged close to 2 million tons during the period 1950–57, were allowed to double to more than 4.1 million tons in 1959. In 1960, at the height of the famine, the government approved net grain exports of more than 2.6 million tons. It was not until 1961 that China responded to the food crisis with massive grain imports. Grain exports were cut in half and imports increased from less than 100,000 tons to more than 5.8 million tons; net imports approached 4.5 million tons. The effect was dramatic—estimated average per-capita energy availability in 1961 (1,763 kcal) was 11.5 percent greater than in 1960 (1,578 kcal), with about two-thirds of the increase resulting from the switch to net grain imports.

The government's failure to respond to the food crisis with large grain imports in 1959–60 is largely attributable to the systematic falsification of production data at the end of the 1950s; government leaders were apparently unaware of the extent of the collapse of agricultural production and the consequential loss of life.[12] Large grain imports in 1961 did greatly increase per capita food availability. However, the government's two-year delay in responding to the food crisis entailed tremendous human costs.

---

[11] Shortfalls of domestic grain production do not explain the increase in grain imports in recent years since grain production has achieved record levels since 1977, increasing far more rapidly than population growth. The dynamics of China's agricultural trade have been reviewed by Frederick Surls, "Foreign Trade and China's Agriculture," in *The Chinese Agricultural Economy*, ed. Randolph Barker and Radha Sinha (Boulder, Colo., 1982), and Riley Kirby, "Agricultural Trade of the People's Republic of China," *U.S. Department of Agriculture, Foreign Agricultural Economic Report No. 83* (Washington, D.C., 1972).

[12] This point of view is proposed by Basil Ashton, Kenneth Hill, Alan Piazza, and Robin Zeitz, "Famine in China, 1958–61," *Population and Development Review* 10 (December 1984): 613–45.

Food Availability and Distribution

Per capita food availability has increased greatly and regional and local disparities in food consumption have been reduced in China since 1949. As a result, the average Chinese now consumes considerably more food than was the case thirty years ago, and the number of malnourished Chinese has declined significantly. It is important to note, however, that food availability has not increased steadily over time nor have all Chinese benefitted equally from the increase. Thus, although the situation is now generally much improved over that of the early 1950s, there have been extreme setbacks along the way, and serious regional and localized shortages of food continue to afflict parts of rural China today.

Food Availability. Trends in national average daily per capita nutrient availability in China for the years 1950–82 are summarized in table 3 and illustrated in figure 1.[13] (Annex table 1 presents year-by-year estimates of per capita nutrient availability with estimates of the fraction available from animal sources.) These figures show that, despite wide fluctuations, per capita nutrient availability has increased markedly and now exceeds estimated nutrient requirements by a large margin. During the thirty years 1950–52 to 1980–82, per capita availability of food energy grew by more than one-third, increasing from about 90 percent to about 115 percent of estimated energy requirements.[14] Per capita availability of protein grew by more than one-quarter, increasing to more than 150 percent of estimated safe levels of protein intake. Per capita availability of fat grew by more than one-half.

The increase in per capita nutrient availability since 1949 has been closely tied to secular trends in China's agricultural production.[15]

---

[13] The estimates of national average per-capita nutrient availability are from annual food balance sheets constructed for China for 1950–82. See Piazza, *Food Consumption and Nutritional Status.*

[14] It is estimated that average per capita food energy requirements and safe levels of protein intake in 1980–82 were about 9 and 10 percent greater, respectively, than in 1950–52. Consequently, the increase in food energy and protein reported in table 3 appears somewhat less when measured as a percentage of requirements than when measured in absolute terms (i.e., in kcal and grams).

[15] For a detailed analysis of agricultural policy and food production in China, see Lardy, *Agriculture in China's Modern Economic Development;* Barker and Sinha, eds., *Chinese Agricultural Economy;* and

Table 2

China's International Grain Trade,
Selected Years, 1950–1965

(- - - - - - - - - in Million Metric Tons - - - - - - - - -)

Year	Total Grain Production[a]	Grain Exports	Grain Imports	Net Grain Imports	Domestic Grain Supply	Per Capita Grain Supply (kg/year)
1950	129.6	1.2	.1	-1.2	128.4	234
1955	180.2	2.2	.2	-2.1	178.1	294
1959	165.2	4.2	.0	-4.2	161.0	246
1960	139.4	2.7	.1	-2.7	136.7	210
1961	143.1	1.4	5.8	4.5	147.6	229
1965	197.5	2.4	6.4	4.0	201.5	282

Source: State Statistical Bureau, *Statistical Yearbook of China: 1983* (Hong Kong, 1983).

[a] The tuber component of total grain production has been included at one-fifth its wet weight for 1950–63.

Table 3

Per Capita Food Energy and Protein Availability and
Requirements, Selected Three-Year Averages

| Period | Energy Availability | | Protein Availability | | Fat Availability |
	kcal/day	As % of Requirement	gm/day	As % of Requirement	gm/day
1950–52	1,894	91	53	133	25
1960–62	1,736	82	46	114	18
1970–72	2,040	97	53	132	25
1980–82	2,570	114	66	151	37

Source: Piazza, *Food Consumption and Nutritional Status.*

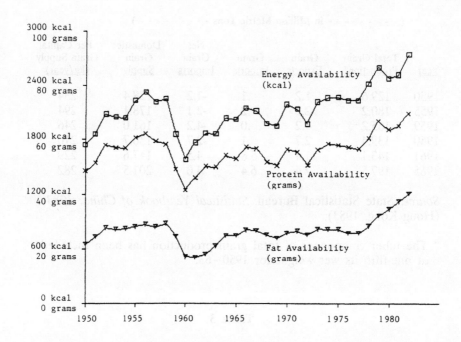

Figure 1. Average Daily Per Capita Nutrient Availability.

Source: Piazza, *Food Consumption and Nutritional Status.*

Agricultural production increased sharply during the early and mid-1950s as China recovered from the devastation of the national and civil wars of the 1940s. Per capita nutrient availability increased rapidly and by the mid-1950s daily per capita availability of food energy topped 2,200 kcal, slightly greater than energy requirements. During the food crisis years of 1959–62, however, agricultural production declined disastrously in response to the chaos created by the "Great Leap Forward" and the sudden implementation of the commune system (1958–59), policy-induced reductions in sown area (1959), and widespread unfavorable weather (1960–61). Per capita nutrient availability fell below the 1950 level and by 1960 daily per capita availability of food energy had declined to less than 1,600 kcal—only 75 percent of energy requirements. The number of deaths due to malnutrition-induced mortality and starvation during the food crisis has been estimated at sixteen million or more.[16] Agricultural production recovered only slowly during the next fifteen years, 1963–77, and it was not until the mid-1970s that per capita nutrient availability regained the level achieved in the mid-1950s. During 1978–84, agricultural production grew at a rapid pace and per capita nutrient availability increased commensurately. Daily per capita availability of food energy and protein now exceeds estimated nutrient requirements by a large margin and compares favorably with the average for middle-income countries. This positive situation is expected to continue for the rest of the century, both because of probable continued production growth and because the rate of population growth has been reduced to less than 1.5 percent per annum.

Since food is unevenly produced and distributed, the nutritional status of the Chinese population may not be directly calculated from these estimates of per capita nutrient availability. Thus, although per capita availability of energy and protein now greatly exceeds requirements for these nutrients, a large segment of the population may still suffer malnutrition. The extent to which a given margin of nutrient

---

Walker, *Food Grain Procurement and Consumption.*

[16] Ansley Coale, in *Rapid Population Change in China, 1952–82* (Washington, D.C., 1984), concluded that the food crisis "led to an excess of about 16 million deaths" (pp. 69–70); and Ashton et al. ("Famine in China, 1958–61") estimated that there were about thirty million extra deaths (over the number that would have been expected on the basis of normal annual mortality rates).

availability over requirements will lead to insufficient food consumption among a portion of the population depends on both the distribution of income and the income elasticity of demand for the nutrient.[17] Although the national average nutrient availability data are only of limited usefulness in assessing the precise magnitude of malnutrition at a given point in time, there appears to be a strong positive relationship between overall nutrient availability and health. Figure 2 illustrates the strong negative relationship between the infant mortality rate (IMR) and per capita food energy availability in China during the period 1950–82. The 1959–62 food crisis caused great human suffering, reflected in a sharp increase in IMR and an actual decline in the total population in two years. The long-term decline in IMR resumed after the food crisis with the recovery of growth in per capita nutrient availability during the last two decades.

Distribution. Increasing average per capita food availability does not in itself reduce the incidence and severity of malnutrition. Malnutrition is significantly curtailed only when an increase in per capita food availability benefits the entire population, not just those already enjoying an adequate diet. For China, the available provincial food production and country- and team-level income distribution data show that the benefits of the increases in per capita food availability since 1949 have been widely distributed. The combination of increased per capita food availability and greater equality of food consumption have together brought about a major improvement in the nutritional status of the Chinese.[18] Although the reduction cannot be accurately quantified, it is clear from evidence presented in the next section that the number of malnourished Chinese has declined both relatively and absolutely.

---

[17] Limited data from China confirm the standard finding that the income elasticity of demand for protein far exceeds that for energy. Hence, the distribution of protein consumption is likely more unequal than that of energy, and it *cannot* be concluded from table 3 that there are likely to be more individuals suffering energy than protein deficiency.

[18] Infectious diseases in children are important causes of malnutrition, and China's advances in controlling infectious diseases have undoubtedly contributed substantially to improved nutritional status. In this paper we do not seek to quantify the relative importance of infectious disease and food availability as causes of malnutrition in China.

Figure 2. Infant Mortality and Per Capita Food Energy.

Sources: Energy Availability: Piazza, *Food Consumption and Nutritional Status.* Infant Mortality Rate: Kenneth Hill, "Demographic Trends in China, 1953–1982," Technical Note 85-4, Population, Health and Nutrition Department, World Bank (Washington, D.C., July 1985).

The comparison of provincial per capita grain production in 1953–57 and 1979–82 in table 4 provides some of the most convincing evidence that food consumption is now more equally distributed in China.[19] Specifically, the figures show that the number of Chinese residing in grain-poor provinces[20] declined from about 220 million in the mid-1950s to about 85 million by the early 1980s. Since total population nearly doubled during the same period, the percentage of the population residing in grain-poor provinces declined by an even greater margin, from about 35 percent in 1953–57 to less than 10 percent in 1979–82. These figures are encouraging, but it should be noted that they also document the emergence of a serious and chronic shortfall in grain production in northwest and southwest China. Thus, although food supplies are adequate at the national level and despite grain transfers from other provinces, provincial food balance sheets for 1979–81 show that per capita availability of energy in Gansu in the northwest and Guizhou, Guangdong, and Yunnan in the southwest was below requirements. The situation in Guizhou appears to have been particularly bleak.

It is also clear that national land reform and the adoption of the commune system greatly reduced rural inequality at the local level (i.e., at the level of the village and the production team) during the 1950s. From 30 to 45 percent of China's arable land and somewhat lesser quantities of other productive assets were redistributed during the national land reform of 1950–52, and this brought about an immediate and substantial reduction in the disparity of wealth and income at the village level.[21] With the formation of the commune system in the mid-1950s, moreover, the system of private ownership of land and other agricultural assets was virtually eliminated in the countryside. Land,

---

[19] Annex table 2 contains average annual data for all provinces on per capita nutrient availability (and trade) for the period 1979–81.

[20] Grain-poor provinces are those with per capita production of unprocessed grain below 275 kilograms per year. Depending upon the grain mix, 275 kilograms of unprocessed grain is considered adequate to meet the subsistence requirements of one peasant for one year.

[21] The national land reform and its effects on the distribution of income are examined in detail in Peter Schran, *The Development of Chinese Agriculture, 1950–1959* (Ithaca, N.Y., 1969), and John Wong, *Land Reform in the People's Republic of China* (New York, 1973).

Table 4

Grain-Poor Provinces, 1953–1957 and 1979–1982[a]

Province	Per Capita Unprocessed Grain Production (kg/year)	Population (Millions)
1953–1957		
Shandong	220	51.5
Henan	240	46.4
Hebei	243	40.4
Qinghai	257	1.9
Liaoning	260	21.3
Shanxi	265	15.1
Jiangsu	274	43.2
Subtotal		220
National	291	610
1979–1982		
Guizhou	229	28.0
Qinghai	230	3.8
Xizang	237	1.9
Gansu	242	19.4
Yunnan	272	32.1
Subtotal		85
National	348	991

Source: Piazza, *Food Consumption and Nutritional Status.*

[a] Provinces producing less than 275 kilograms of unprocessed grain per capita per year are considered to be grain-poor. No data were available for Xizang for 1953–57.

tools, and farm animals were jointly owned by production team members, and, consequently, income disparity *within* the production team was greatly reduced.

It is important to note, on the other hand, that income disparity *between* communes and counties was not affected by the national land reform or the formation of communes. Consequently, substantial differences in average household income between these larger administrative units have remained largely intact.[22] Sharp curtailment of internal migration has further helped maintain geographical pockets of poverty. However, it is known that the incidence of extreme rural poverty has diminshed greatly in recent years. National surveys of household income have documented that, with the increase in agricultural production since 1978, the number of poor counties and production teams has declined tremendously. During 1977–81, the number of poor counties declined by 60 percent and the percentage of poor production teams declined from 39 percent to less than 20 percent of all teams.[23]

Nutritional Status: Secular Trends and the Current Situation

Improvements in Nutritional Status

This section summarizes the anthropometric data documenting the secular improvement in nutritional status in China in the post-1949 period. The data show gains in nutritional status most clearly for the urban population. There are comparatively few rural data, and what data are available are principally for very well-off rural areas (*viz.*, the outskirts of Beijing and Shanghai). Since the available evidence suggests

---

[22] Nicholas Lardy, in *Agriculture in China's Modern Economic Development,* has presented convincing evidence that pockets of chronic rural poverty have continued to exist in several provinces that have experienced tremendous overall agricultural growth since 1949. He also concludes that government policies may have contributed to the impoverishment of such specialized groups as cotton farmers in Shandong, cash-crop farmers in Yunnan, and pasturalists in Gansu during the 1960s and 1970s.

[23] A "poor" county or team is defined as one with a per capita income of less than fifty *yuan* per annum. See Lardy, *Agriculture in China's Modern Economic Development,* 171–72.

that the secular improvement in nutritional status in these rural areas has at least equalled the gain in urban areas, however, the urban data probably provide a reasonable proxy for trends in nutritional status of the general population.[24] Since the data also show that nutritional status in rural areas lags far behind that in urban areas, on the other hand, the urban data overstate the absolute level of nutritional status of the general population. Current nutritional status and regional and rural-urban differences in it are discussed below.

Secular Improvements in Nutritional Status. The data from several anthropometric surveys undertaken during the 1950s and in 1979 make possible an assessment of secular trends in the nutritional status of the Chinese. It has already been noted that these anthropometric data are for urban China and therefore overstate the nutritional status of the general population. It is also known that the data overstate nutritional status since the surveys exclusively measured students, and students are uniformly taller than the general population.[25] Another concern is whether the specific years for which data are available represent average years in terms of agricultural production and general food availability, since short-term food shortages or surpluses in the survey year could be reflected in the weights of children. However, it is far less likely that heights would be markedly affected by short-term food availability, so that height measurements may provide a particularly useful indication of longer-term growth trends. Keeping these limitations in mind, the

---

[24] Data for Beijing and Shanghai municipalities show that the anthropometric status of their rural populations has increased more rapidly than has that of their urban populations. In Anhui Province, rural height increased by 4.2 centimeters per decade during the period 1958–79. The increase in urban height, on the other hand, was only 2.3 centimeters per decade (see Piazza, *Food Consumption and Nutritional Status*). It should be stressed, however, that the available rural anthropometric data represent limited samples from well-off areas. Data on trends in anthropometric data represent limited samples from well-off areas. Data on trends in anthropometric status in poor rural areas are unavailable (although current levels are known in some areas and are quite low).

[25] J. M. Tanner, *Foetus into Man: Physical Growth from Conception to Maturity* (London, 1978), 149, discusses the difference in anthropometric status between the student and the general population.

available anthropometric data for China are nevertheless useful for assessing secular change in nutritional status.

National average height and weight for boys of several ages in China in 1979, 1951–58, and 1915–25 are summarized in table 5. The national averages are calculated from anthropometric surveys of sixteen provinces conducted in 1979, fifteen provinces in 1951–58, and three provinces in 1915–25.[26] These figures document tremendous improvement in urban nutritional status during the twenty-five-year period from 1951–58 to 1979. Male and female height and weight in 1979 exceeds that of 1951–58 by a large margin for all ages. Males between the ages of seven and twenty in 1979 were on average 7.6 centimeters taller and 4.3 kilograms heavier than males of the same ages in 1951–58. Females in 1979 were on average 6.6 centimeters taller and 3 kilograms heavier than were females of the same ages in 1951–58. These differences correspond to an average increase in height and weight of about 1.2 and 0.6 standard deviations, respectively.[27]

Data on secular change in anthropometric status are often reported as growth rate per decade. Tanner reports, for example, that European children of average economic circumstances have increased in height at ages five to seven years by between 1 and 2 centimeters per decade since 1900 and that Japanese seven-year-olds increased by the larger

---

[26] The sixteen provinces and municipalities covered by the 1979 survey were Beijing, Shanghai, Tianjin, Shandong, Shanxi, Shaanxi, Gansu, Liaoning, Heilongjiang, Anhui, Fujian, Guangdong, Yunnan, Sichuan, Hunan, and Hubei. See Research Group, *Research on the Physical Characteristics, Fitness and Vital Indicators of Chinese Children and Young Adults* (Beijing, Science and Technical Papers Publishing House, 1982). The national average for the 1950s includes data from Beijing, Shanghai, Shanxi, Shaanxi, Liaoning, Jilin, Heilongjiang, Henan, Jiangsu, Guangdong, Guangxi, Sichuan, Hunan, and Hubei. See Piazza, *Food Consumption and Nutritional Status*, 212, n. 12, for complete references. The 1915–25 national average corresponds to data primarily from Shandong, Jiangsu, and Guangdong. See Paul Stevenson, "Collected Anthropometric Data on the Chinese," *Chinese Medical Journal* 39 (October 1925): 855–98.

[27] The improvement in national average nutritional status corresponds closely to provincial improvements over the same period reported by Jamison and Trowbridge, "Nutritional Status of Children in China," and by Piazza, *Food Consumption and Nutritional Status*.

Table 5

Heights and Weights of Male Children, 1915–1925, 1951–1958, and 1979

Year	Height (cm)			Weight (kg)		
	Age 7	Age 12	Age 18	Age 7	Age 12	Age 18
1915–25	114.6	136.3	163.1	18.4	29.4	50.1
1951–58	114.5	135.7	163.6	19.8	29.2	51.6
1979	120.7	144.7	168.9	21.1	33.6	56.2
Decennial Improvement 1951–58 to 1979	2.6 cm	3.8 cm	2.2 cm	0.5 kg	1.8 kg	1.9 kg

Source: Piazza, *Food Consumption and Nutritional Status.*

margin of about 3 centimeters per decade during 1950–70.[28] By comparison, the national average increase in height for Chinese seven-year-olds from 1951–58 to 1979 was about 2.6 centimeters per decade. Thus, the decennial increase in height for age in China surpassed that of many European countries during the twentieth century and almost equalled that of Japan during its most rapid period of anthropometric improvement. If the European experience is any guide, moreover, these secular increases can be expected to continue well into the next century. Tanner also observes that conditions that lead to secular increase in height have two effects—to increase the ultimately attained height and to increase the rate of maturation. The increased maturation rate results in higher rates of secular change in height in youth (ages ten to fourteen years) than in young adults. This tendency is clearly evident in the bottom row of table 5.

The data for 1915–25 are not as useful for secular comparisons since they are somewhat more positively biased than the data for the 1950s and 1979.[29] Nevertheless, the contrast between the rapid improvement in nutritional status from 1951–58 to 1979 and the lack of any significant change in nutritional status from 1915–25 to 1951–58 is quite striking. The post-1949 improvement in nutritional status is most importantly explained by the increase in per capita food availability and its more equal distribution and improvements in health relative to the pre-1949 period. It is interesting to note that since per capita food energy and protein availability did not increase during the period 1958–77 (fig.1), the improvement in nutritional status during this period does not appear to be related to changes in nutrient availability. This suggests that improvements in health—principally, the reduction of diarrheal infections—have played a key role. Another possibility is that the distribution of food has changed in favor of the relatively well-off areas, from which most of the anthropometric data come. The national

---

[28] Tanner, *Foetus into Man*, 150–51.

[29] Most of the anthropometric surveys undertaken in China are positively biased since they only measured healthy students residing in urban areas. The 1915–25 survey was more biased than were later surveys since students were a smaller minority of the total population in the early twentieth century. It also seems likely that the difference between the student and the general population was greater in 1915–25 relative to the post-1949 period.

land reform carried out during 1950–52 may, in addition, have created preconditions for subsequent improvements in nutritional status by greatly reducing intraregional disparities in capacity to produce food. This, combined with protection of the very poorest against complete food deprivation, may have contributed substantially to the reduction in malnutrition-related child deaths from 1950 onwards.

Nutritional Status Today

The adoption and increasing implementation of a one-child family policy in China has generated strong concern that each child have full opportunity for optimal physical, mental, and psychological development. Prenatal nutrition and counseling for expectant mothers is one important facet of policies to improve child development, and the empirical evidence, although highly selective, does suggest that the prevalence of low birthweight (less than 2.5 kilograms) is low. China's success with fertility limitation has also almost certainly had important benefits for the health of children actually born. Children also seem well protected against vaccine-preventable diseases, and, although there remain problems of diarrheal, respiratory, and parasitic infections, substantial progress has been made, in much of China, in transforming these from major sources of mortality to lingering problems of morbidity. Progress against malnutrition has been so substantial in urban areas that it can no longer be considered a problem of child development. Rural areas, in contrast, continue to suffer substantial amounts of moderate-to-serious child malnutrition.[30]

Rural-Urban Differences. China's 1979 anthropometric survey of sixteen provinces and municipalities provides one valuable data set for a

---

[30] See A. Berg, *Malnourished People: A Policy View* (Washington, D.C., 1981), 9–15, for an overview and references to the literature concerning adverse functional consequences of malnutrition. Data from three province-level units in China—Beijing, Gansu, and Jiangsu—have also been analyzed to ascertain the effect of malnutrition on schoolchildren's performance in those areas; even after controlling for whether the child was in a rural or an urban school, low height-for-age was consistently found to affect performance adversely (as measured by the number of grades behind a child was in comparison to where he should be, given his age). See D. Jamison, "Child Malnutrition and School Retardation in China," *Journal of Development Economics* 20 (1986): 299–309.

careful assessment of urban-rural and interprovincial differences in child malnutrition. A second data set, from 1975, provides information on a more limited sample—nine cities (three each in northern, central, and southern China) and the suburban areas immediately outside them.[31] A clear pattern of rural malnutriton can be seen in these data, and relatively little urban malnutrition. Table 6 reports both the percentage of seven-year-old children stunted (i.e., unduly low in height-for-age) and the percentage of those low in weight-for-age (in the 1979 survey). Two points emerge from the table. First, rural-urban differences are marked, although somewhat less so in weight-for-age than in height-for-age.[32] In the most favored locations (urban Beijing and Tianjin), there appears to be virtually no stunting. Secondly, both urban and rural malnutrition vary quite considerably from province to province. About 37 percent of rural boys in Sichuan are stunted, in contrast to only 12 percent in rural Shandong and less than 4 percent in rural Tianjin. (Although the corresponding data for females are not presented in the table, it should be noted that there did not appear to be any important differences between males and females in percent stunted.)

Sources of Differences. One potential source of the differences in prevalence of malnutrition reported in table 6 is variation in genetic potential between northern and southern Chinese. However, the relatively small differences between cities in northern and southern provinces (compare Beijing and Guangdong) suggest that genetic differences are relatively unimportant; and they certainly could not explain rural-urban differences in the same province. A second and well-established source of variation in growth is variation in disease prevalence, particularly diarrheal disease prevalence. Although the hard evidence is scant, diarrheal disease almost certainly remains an important problem in rural China, and hygienic improvements that reduce

---

[31] Annex tables to Jamison and Trowbridge, "Nutritional Status of Children in China," excerpt results from the data published from these surveys and transform those data into various standards utilized for making comparisons across different groups (male-female, age groups, provinces, etc.).

[32] Available data consistently show rural Chinese children to be as high (often higher) in *weight-for-height* as their urban counterparts. Greater stockiness of rural children may, then, be partially compensating for lower stature in reducing rural-urban differences in weight-for-age.

Table 6

Percentage of Seven-Year-Old Boys Malnourished, Selected Provinces, 1979

Province/Municipality	% Stunted[a]		% Low Weight for Age[b]	
	Urban	Rural	Urban	Rural
Beijing	0.7	8.4	3.8	10.2
Tianjin	0.5	3.8	2.9	4.4
Shandong	2.1	11.9	6.2	11.3
Heilongjiang	1.3	19.5	5.9	15.4
Guangdong	1.3	19.2	6.4	23.0
Sichuan	7.5	37.1	11.1	26.4
National average	2.6	12.7	7.6	13.1

Source: D. T. Jamison and F. Trowbridge, "The Nutritional Status of Children in China: A Review of the Anthropometric Evidence," Technical Note GEN 17, Population, Health and Nutrition Department, World Bank (Washington, D.C., 1984).

[a] A child is said to be "stunted" if his or her height is less than 90 percent of the median height of children of the same age by the standards compiled by the U.S. National Center for Statistics (NCHS). The percentage stunted can be calculated from the mean and standard deviation of the distribution of height at a given age, assuming the distribution to be normal, which is a reasonable assumption.

[b] The percentage of those low in weight for age was calculated assuming a normal distribution for weight (which is not a particularly good approximation) and defines low weight for age as below 75 percent of the NCHS median. This corresponds to either Gomez II malnutrition (60–75 percent of NCHS median) or Gomez III malnutrition (below 60 percent of NCHS median).

diarrheal incidence would likely have beneficial nutritional effects. Finally, differences in the quantity and quality of available food probably account for much of the rural-urban difference in prevalence of malnutrition. The State Statistical Bureau reports that in 1978 the average urban resident consumed 22 percent more food energy, 14 percent more protein, and 90 percent more fat than his rural counterpart. Although the difference was somewhat decreased in 1982–83, it still amounted to about 400 kcal of energy and about 40 grams of protein and fat per capita per day.[33]

Efforts to reduce rural malnutrition will, therefore, need to address very directly the problem of providing more food, and possibly food of higher protein content, to rural children and youth. Diarrheal disease control efforts may also be important, but probably relatively less so than in many other countries. The increases in rural income that have ensued from economic reform in recent years will undoubtedly be important for many areas in providing the food required to reduce malnutrition. However, close surveillance of children's growth to allow targeting of food to the malnourished is a contribution the health care system can make to reducing rural malnutrition, and, where it is not now done, this effort should be a priority of the maternal and child health system.

Micronutrient Deficiency Diseases. Generally speaking, micronutrient deficiencies appear to be less of an issue in China than in many low-income countries. However, mild anemia is widespread, in part because of inadequate iron intake in the diet and in part because of iron losses due to hookworm infestation. (In southern China hookworm infection appears to be highly endemic, but the problem appears to receive little

---

[33] See State Statistical Bureau, *Statistical Yearbook of China: 1984* (Hong Kong, 1984), 480. Urban food subsidies amounted to approximately 96 *yuan* per capita in 1981; there are virtually no food subsidies in rural areas (aside from foregone interest on state loans for food to communes producing below the ration level). If the urban food price subsidy is added to per capita expenditures on food, urban food expenditures total 354 *yuan* per capita per annum, in contrast to rural expenditures of 114 *yuan*. The state subsidy accounts for 40 percent of the difference between rural and urban food expenditures. Estimates of rural and urban food expenditures for 1981 appear in the *Statistical Yearbook of China: 1981* (Hong Kong, 1981), 439 and 445.

attention from health authorities.) Other deficiency diseases that remain important in China are rickets (vitamin D deficiency), goiter (iodine deficiency), and Keshan disease (selenium deficiency). There is relatively little vitamin A deficiency, which, in a number of other countries, is an important cause of impaired vision and perhaps of higher mortality rates. Control of goiter has been the subject of extensive campaigns in recent years, and if these efforts continue at an adequate level, goiter could cease to be an important problem in the near future. Keshan disease seems to be found exclusively in perhaps fifteen provinces in China; it is virtually unknown elsewhere in the world. It frequently leads to death through heart failure among children or among women of childbearing age. Almost half of nutrition research in China is directed to understanding this disorder, which is now relatively unimportant as a public health problem. Selenium supplementation of food in affected locales seems effective in reducing incidence.

Issues for the Future

The Chinese Government has been active in setting food policy and is to be credited with both major failure and major success in this regard. The most notable failure was the policy-induced famine of 1959–61, which caused the premature loss of perhaps thirty million lives.[34] Net exports of grain were allowed to double during the first two years of the famine, contributing to this tragedy. The government may also be faulted for having brought about reduced production of soybeans, other pulses, and oilseeds during the 1960s and 1970s. This action further diminished protein complementarity and fat availability in the average diet. On the positive side, government policy has been instrumental in achieving a long-term increase in per capita food availability and reducing local disparities in food consumption. The rationing system contributed to the virtual elimination of malnutrition from major urban areas and grain transfers have moderated regional shortfalls in food production. Until recently, the government had also constrained the growth of meat consumption, and this has had the positive effect of making grain available to China's poor at affordable prices. In sum,

---

[34] For an analysis of data concerning the famine, see Ashton et al., "Famine in China, 1958–61."

although there have been serious errors along the way, food policy has played an important and generally positive role in reducing the incidence and severity of hunger and malnutrition in China since 1949.

As a result of the success of these policies, China's food problem is no longer one of insufficient national production. To the contrary, per capita nutrient availability in China now equals that of the average middle-income country (table 1). The remaining nutrition problems are to guarantee adequate levels of food intake among the poor and to avoid problems of over-nutrition, with its concomitant enhanced risk of heart disease, stroke, and some cancers.

At least three issues arise concerning China's efforts to reduce remaining malnutrition. First, the centerpiece of recent agricultural reform in China, the "production responsibility system" and related reforms of the commune system, may prove inimical to the welfare of indigent households and individuals.[35] These reforms emphasize the link between income and individual work effort and have been central to the revitalization of agricultural incentives. However, since households with few or no able-bodied workers participate in agricultural production to only a limited extent, they have not directly benefitted from the resultant increase in production in recent years. Such indigent households are dependent upon local relief programs for food, medical care, education, and other basic needs.[36] Unfortunately, the ongoing reform of the commune system (which includes the elimination of communes as political organizations) may reduce or disrupt such welfare activities, at least in the short run.[37]

---

[35] The production responsibility system and other recent agricultural reforms are discussed in Azizur R. Khan and Eddy Lee, *Agricultural Policies and Institutions after Mao* (Singapore, 1983).

[36] John Dixon, in *The Chinese Welfare System, 1949–1979* (New York, 1981), presents a detailed review of China's welfare system.

[37] There have been numerous reports of production teams and brigades reducing funding for and even abandoning their local health clinics in order to increase the personl income of team and brigade members. Similarly, cadres are being extolled to maintain the social welfare system (the "five guarantees")—a sure sign that the welfare system has been undermined or discontinued in some areas. It is important to note, however, that the available national data show no indication of an increase in the number of the extremely poor. Such data suggest, in fact, that increased income among large numbers of

Second, the government has committed itself to ambitious plans to double per capita consumption of red meat and other preferred foods by the year 2000.[38] Large increases in the production of preferred foods need not necessarily bring about a reduction in the amount of grain available for direct consumption by China's poor, since national grain supplies more than meet current demand, and grain production is expected to continue to increase more rapidly than population. However, upward pressures on grain prices, should they occur, could only be expected to have a particularly adverse effect upon the poor. Furthermore, the task of increasing meat production for the majority of Chinese may distract the government's attention from the task of increasing grain consumption for the malnourished. Unfortunately, guaranteeing adequate food intake among China's poor is both a less tractable and less politically rewarding endeavor than that of increasing the consumption of preferred foods. It would, however, represent a far more important contribution to nutritional status in China.

A final issue concerning food and nutrition policy in China involves the medical consequences of the transition that China has embarked upon toward satisfying a much higher percentage of its dietary requirements from animal sources. Reductions in mortality due to infectious disease and undernutrition have brought China a profile of health problems—stroke, cancer and heart disease—remarkably similar to those of industrialized countries. While definitive conclusions regarding the relation between chronic disease and excess consumption of animal products cannot yet be drawn, the available epidemiological evidence strongly suggests that, if the Chinese substantially increase their consumption of animal products, particularly of animal fat, this will result in a quantitatively important increase in mortality from heart disease, stroke and some cancers. Substantial taxes on animal consumption—and, possibly, production controls—could be expected to help China avoid key problems of overnutrition, improve state revenues and foreign trade, and maintain lower grain prices.

---

poor households have more than offset the possible loss of income among some indigent households with high dependency ratios.

[38] Wu Daxin summarizes China's agricultural production targets for 1990 and 2000 in "Briefing on Agriculture of the PRC," in *Agricultura Sinica*, ed. E.M. Reisch (Berlin, 1982), 13–25.

Annex Table 1

Per Capita Nutrient Availability in China, 1950–1982

Year	Per Capita Grain Production (kg/year)	National Average Daily per Capita Nutrient Availability			Nutrient Availability from Animal Sources		
		Energy (kcal)	Protein (gm)	Fat (gm)	Energy %	Protein %	Fat %
1950	236	1,742	49	22	4	6	25
1951	251	1,856	51	25	4	6	25
1952	281	2,083	58	28	4	6	26
1953	280	2,048	57	27	4	7	27
1954	279	2,041	57	28	4	7	28
1955	297	2,232	61	28	3	7	24
1956	304	2,326	62	29	3	6	22
1957	301	2,217	59	28	4	7	27
1958	299	2,248	58	29	4	8	29
1959	253	1,854	49	25	4	8	28
1960	214	1,578	41	17	4	8	28
1961	222	1,763	46	17	3	6	23
1962	238	1,867	50	18	3	6	22
1963	246	1,857	50	21	4	7	30
1964	269	2,037	54	25	4	8	31
1965	272	2,021	53	25	5	8	35
1966	291	2,154	57	27	5	8	34
1967	288	2,118	57	27	5	8	34
1968	269	1,979	52	25	5	8	34
1969	264	1,950	50	24	5	8	34
1970	292	2,192	56	26	4	7	32
1971	297	2,138	55	27	5	8	35
1972	279	1,972	51	25	6	10	40
1973	300	2,219	56	27	5	9	36
1974	305	2,264	58	27	5	8	35
1975	310	2,266	58	27	5	8	36
1976	307	2,235	57	26	5	8	36
1977	299	2,233	56	26	5	8	36
1978	318	2,413	60	28	5	8	35
1979	342	2,592	67	32	5	8	37
1980	326	2,473	64	35	6	9	38
1981	327	2,511	65	37	6	9	37
1982	351	2,725	70	40	6	9	36

Source: Piazza, *Food Consumption and Nutritional Status*.

Annex Table 2

Per Capita Nutrient Availability by Province, 1979–1981 Average

	Energy		Protein		Average Annual Grain Imports (kg per capita)[a]
	Availability (kcal/day)	% of Requirements	Availability (grams/day)	% of Safe Level of Intake	
Hebei	2,575	121	67	166	8
Shanxi	2,419	114	62	156	14
Inner Mongolia	2,311	109	58	146	25
Liaoning	2,874	135	76	188	37
Jilin	3,134	147	83	202	Exporter
Heilongjiang	3,134	147	99	239	-17
Jiangsu	2,857	134	65	162	-17
Zhejiang	2,656	125	56	140	Exporter
Anhui	2,228	105	50	128	-16
Fujian	2,316	110	45	116	4
Jiangxi	2,303	109	41	106	-19
Shandong	2,774	130	72	179	No trade
Henan	2,319	110	62	155	No trade
Hubei	2,393	113	50	128	-8
Hunan	2,402	113	44	112	-13
Guangdong	2,093	99	40	103	6
Guangxi	2,154	102	40	103	Importer
Sichuan	2,296	109	50	126	No trade
Guizhou	1,637	78	36	95	22
Yunnan	1,928	92	43	110	10
Xizang	2,269	107	68	169	N/A
Shaanxi	2,325	110	60	150	21
Gansu	2,094	99	55	139	31
Qinghai	2,498	118	67	165	54
Ningxia	2,522	119	64	159	19
Xinjiang	2,740	129	73	179	18

Source: Piazza, *Food Consumption and Nutritional Status.*

[a] These figures are adopted from Kenneth Walker, *Food Grain Procurement and Consumption in China* (Cambridge, England, 1984), 187–89. It is assumed that Chinese sources have reported grain trade in "trade grain," i.e., husked rice and millet, but all other grains in their unprocessed form. Assuming 50 percent husked rice and 50 percent wheat flour, each kilogram of grain would provide 3,350 kcal of energy and 81 grams of protein. Jilin and Zhejiang are known to be net exporters of grain and Guangxi to be a net importer, but actual quantities are not known. No trade data were available for Xizang.

World Bank Experience in Education in China, 1980–1984

Frank Farner

China had long been a member of the World Bank when, in 1980, the People's Republic assumed the seat on the World Bank membership board that had previously been occupied by the Republic of China on Taiwan. To give proper acknowledgement of this momentous change, a high-level delegation from the World Bank, consisting of vice presidents and department directors from the leading sectors (agriculture, transportation, and energy), assembled in Beijing. The Chinese immediately asked, "Do you enter the education field?" (No representative from this sector had been included in the delegation.) The leader of the Bank delegation replied, "Yes, we do, but we didn't think you would want any of that." When the Chinese indicated that they were indeed interested in talking about education, a World Bank education officer was dispatched to China to begin discussions on the Chinese education program, with a particular focus on higher education.

From 1949 to 1980, primary education enrollment ratios in China increased from 23 percent to 93 percent, rising far above the levels obtaining in other developing countries. Similar advances were made at the secondary level during this period, with enrollment ratios rising from 2 percent to 47 percent. Despite these gains in primary and secondary education, the enrollment ratio for higher education remained far below that of other developing nations, which rose to a level of about 5 percent—and in the case of India, as high as 10 percent—while China followed a very erratic pattern of going up and down without ever attaining similar success. China's level of 1 percent in 1984 remained far below target, representing only about one million students enrolled in roughly eight hundred institutions of higher education. These figures do not include students of the growing television university and do not reflect the development of the last several years, to which we now turn.

In 1980, China graduated eighty thousand students with bachelor's degrees; of these, only some thirty thousand were in the fields of science and engineering. (When the Chinese asked the World Bank for help in education, and specifically higher education, they meant science and engineering.) The first joint project was designed to provide assistance, in the form of a $200 million[*] loan/credit, to twenty-six leading universities. The Chinese Government agreed to contribute another $100 million, for a total of $300 million.

Roughly four-fifths of the World Bank funds were earmarked for laboratory equipment and one-fifth for "staff development," a term used to denote fellowships to enable faculty of the twenty-six universities to study abroad and to send experts to China for the purpose of improving programs of instruction. All of the money provided by the Chinese Government was allocated to the construction of buildings to receive the new lab equipment. The government originally wanted all of the World Bank funding to be used for equipment, but as the talks progressed, the figure of about 20 percent for staff development was eventually arrived at. The bank viewed this allocation as essential, since there would otherwise be a shortage of staff trained to use the new equipment. The project has provided funding for some 1,200 fellowships for Chinese scholars to study abroad, around 800 of whom have travelled to the United States, a figure that the World Bank, for political reasons, would rather was lower. Roughly 350 foreign experts have gone to China under this program, of whom about half were from the United States.

The World Bank had hoped in this first project to realize both a quantitative expansion and a qualitative improvement, but it was not possible to achieve any dramatic increases in enrollments. An increase of 23,000 students (from 92,000 to 115,000) was expected between 1979 and 1985, but this is not very significant when spread over six years and the twenty-six universities that were included in the project, most of

---

[*] All money amounts cited in this essay are in U.S. dollars.

which are relatively small. The principal obstacle to enrollment expansion has been the Chinese insistence that nearly all students reside in dormitories for all four years, with the result that dormitory space has become the major limiting factor. So the bank has had to settle for only about a 25 percent increase in enrollments and concentrate its effort primarily on improvements in the quality of education.

In 1984, the student-faculty ratio in these universities was 3.5:1. This has been a major issue at the World Bank, where it is believed that such ratios are very inefficient and ratios of 8:1 or 10:1 are advocated. More reasonable proportions, of course, become a matter of bargaining between the two sides.

Hopes for qualitative improvements have had a better chance of success, centered largely around the expectation that new buildings will be better designed to accommodate improved curricula. The World Bank strongly believes that the new equipment going to the twenty-six universities—80 percent of which had been delivered by mid-1984—will provide the impetus for such improvements. The nearly one thousand scholars who by then were studying abroad will, it is hoped, return to China to make good use of their training, although there is some concern that many will not have the chance to teach or do research in their areas of specialization for many years, until older professors retire or die.

It is further hoped that the unique foreign expert component of the project, which is overseen by an advisory panel, will direct foreign specialists to departments within universities rather than to advisory positions in the ministries. On the other hand, the reports of the first thirty foreign specialists who went to China have revealed lower faculty quality in the leading universities than might have been expected.

China's continued interest in borrowing for education has been noteworthy. Only a year after inception of the first project, the World Bank extended a $75 million credit to the Ministry of Agriculture for improvements in the eleven leading agricultural colleges and in eight agricultural research stations. This project employed nearly the same design as its predecessor, with about 80 percent of the funding designated for equipment purchases and 20 percent for staff development.

In the summer of 1982, the Ministry of Education requested funding for a project to establish a system of polytechnical colleges at the postsecondary level and to expand the central radio and television university and the provincial television universities. This project has much more potential for quantitative expansion because it avoided the dormitory requirements that restrict enrollments in the leading universities. Seventeen polytechnics have been established so far. These schools have somewhat higher standards than American junior colleges and a selective admissions policy. Three very important new policies are in effect at these polytechnics: (1) the schools are completely nonresidential; (2) students pay tuition, which was previously unheard of; and (3) graduates are not placed by the schools but are left to find jobs on their own. The lack of a placement service may seem a negative aspect for the student, but the application rate for the polytechnics is, in fact, higher than for many of the less well-known full-time universities, because polytechnic students know they will not be posted anywhere except to their own city after graduation. Enrollments in the polytechnics have already exceeded targets for 1988, and students with very high admissions scores are frequently choosing the polytechnics over longer-established institutions. In the television university system, enrollments similarly overtook the original projection of roughly one million students (equal to the whole system of convential residential colleges and universities).

In 1984, the Ministry of Agriculture received World Bank assistance for a second project very similar to its first, although this second project includes two components that we hope will continue in the future. One is the provision of assistance to agricultural secondary schools (an area in which visiting agricultural specialists see a great need); and the other is assistance to the radio program for farmers mentioned in Professor Metcalf's paper. The Ministry of Education also asked for a second university development project, which has already been appraised and will be active within a year. The education projects, then, assisted by the World Bank in 1981–1984, including the project in medical education on which John Evans worked and a proposed agricultural research project, total about $700 million in bank assistance, with the Chinese Government's share approaching at least another $2 billion.

Three new projects in education—intended to effect improvements in teacher education, secondary vocational schools, and instructional materials—are currently in the works and will require an additional investment on the part of the bank of about $300 million. Thus, World Bank assistance to China, the bulk of which is for higher education, will by 1988 total some $1 billion, with roughly double that sum being contributed by the Chinese Government.

The goal of these efforts is to increase enrollments in higher education to about ten million, a substantial jump from the present level of two million (of whom about one-half are now enrolled in regular university programs and the other half in the television university). Such an increase would mean that China would have roughly the same proportion of young adults in higher education as India and would be able to meet most of the needs of its rapidly changing economy.

On the qualitative side, which, of course, is ultimately more important than the quantitative, these projects are being engineered to achieve improved curricula in all fields but especially in science and engineering and in economics and management. Throughout the government there is an awareness that management is one of China's weak points.

There will soon be increased opportunities and incentives for Chinese universities to offer graduate degrees. In the past, these universities offered about one hundred master's degree programs but no doctoral programs. The World Bank is very much involved in the establishment of programs at both levels. A delegation of members of the Council of Graduate Schools, comprised of sixteen graduate deans from around the United States, recently toured China at that government's request.

Research is another area in which major improvements are hoped for, since Chinese professors are continuing to do research in some unproductive fields or in areas that have already been researched, their work often duplicating studies done in the 1950s. Also on the list of objectives are improved utilization of staff, space, and time, and increasing student-faculty ratios from 3.5:1 to 8:1.

It will be instructive over the next decade to compare the renaissance in Chinese education supported by the Rockefeller Foundation during the 1920s and 1930s with what the World Bank is doing today. The reader will note that nearly all of the efforts described above have been phrased in the form of hopes and aspirations, objectives and targets. We at the World Bank await the outcome of these projects with great optimism.

Medical Education in China

John R. Evans[*]

During the thirty-five-year existence of the People's Republic of China, the health status of its citizens has improved more rapidly than that of other low-income countries, approaching the levels found in industrialized countries that devote vastly greater financial resources to health services (table 1). The 1980 health indicators for one of the most advanced areas, Shanghai County, are very close to those of the United States and Canada (table 2). The higher infant mortality in national figures is a reminder that substantial regional disparities in health status persist. The disease profile of the leading urban and rural areas is now typical of an industrialized country, with stroke, cancer, heart disease, and chronic respiratory disease as the leading causes of death (table 3). Health has been a consistent priority for the People's Republic of China, and current policy in the era of the Four Modernizations is no exception. Prime targets for technological upgrading are the educational and research programs of medical colleges and hospital-based health services.

---

[*] The views and interpretation in this article are those of the author and should not be attributed to the Ministry of Public Health, People's Republic of China, or to the World Bank, its affiliated organizations, or anyone acting on their behalf.

Table 1

Indicators Related to Health Status

	Life Expectancy at Birth (in Years)		Infant Mortality Rate (per 100 Live Births)		Child Death Rate (per 1,000 Aged 1–4)	
	1960	1981	1969	1981	1960	1981
21 low-income countries	41	50	163	124	30	21
India	43	52	165	121	26	17
China	41	67	165	71	26	7
19 industrialized countries	70	75	30	11	2	1

Source: World Development Report 1983, World Bank (Oxford, 1983), 192–93.

Table 2

Health Statistics of Shanghai County, 1980
(11 Million Population)

Life expectancy at birth (years)	72.5
Infant mortality rate (per 1,000 live births)	15.8
Child mortality rate (per 1,000 aged 1–4)	1.3
Crude birth rate (per 1,000)	14.8
Crude death rate (per 1,000)	6.2

Source: Xing-yuan Gu and Mai-ling Chen, "Health Services in Shanghai County: Vital Statistics," *American Journal of Public Health* 72 (1982): 19–23.

Table 3

China: Principal Causes of Death, 1980
(Death Rate per 100,000)

	Urban (17 Cities)	Rural (38 Counties)
Cerebrovascular diseases	135	113
Heart diseases	133	171
Malignant tumors	115	97
Respiratory diseases	52	79
Gastrointestinal diseases	23	35
Trauma	19	18
Tuberculosis	12	21
"Toxicosis"	10	18

Source: The Health Sector in China, World Bank Report no. 4664-CHA (Washington, D.C., 1984).

The first school of Western medicine in China, the International Medical School in Guangzhou (Canton), was founded by missionaries in 1866, but Western medicine did not begin to be introduced systematically until 1917. By 1949, there were fifty-six medical colleges and faculties in China, operated more or less independently by missionaries, foreign organizations, or the central and provincial governments. Peking Union Medical College, established in 1924, holds a special place in this period of medical education. Although only 313 doctors were graduated in the college's twenty-four years of operation, these graduates rose to positions of leadership and exercised a profound impact on medical education and scientific research in China.

Most of the medical colleges were located in large urban areas, and since the total number of graduates averaged only five hundred per year, access to Western medical services was limited to a very small segment of the population. The vast majority of Chinese, particularly the rural population, received medical care from practitioners of traditional medicine, who had perpetuated their art by apprenticeship for more than two thousand years. Secondary medical schools that trained nurses and midwives graduated almost 40,000 health workers up to 1949.

In the reconstruction of China that followed the change of government in 1949, an ambitious program was launched to expand and upgrade the training of health personnel. Small colleges were amalgamated, and some were relocated to achieve better geographic distribution of trained personnel. Efforts were made to emphasize preventive measures and to integrate traditional Chinese medicine and Western medicine. By 1957, the number of medical colleges, consolidated to 22 in 1949, had increased to 37, formal training in traditional Chinese medicine had been initiated in 5 new colleges, and training of paramedical workers, assistant doctors, and nurses had expanded more than fivefold in 182 secondary medical schools. With the Great Leap Forward in 1958, the pace of expansion accelerated: new institutions were established, enrollments increased rapidly, academic standards were relaxed, and part-time medical education was encouraged. This surge of activity overextended the resources of the educational institutions and led to a period of retrenchment and consolidation in the early 1960s.

From 1966 to 1975, medical education in China was a victim of the general disruption that resulted from the Cultural Revolution. Initially, colleges were closed, but students continued to graduate on schedule even though they attended no classes. When the colleges reopened, a course of three years became the norm for medical degree candidates. The selection of students and the curriculum were politicized, and admission standards were reduced to lower middle school training and selection based on non-academic criteria. The academic quality of the medical colleges was seriously impaired by a set of policies that sent teachers and hospital doctors to the countryside for extended periods, stopped all research, converted laboratories to production, and dissipated library resources. The consequences in medicine, as in other sectors, were a decade of medical graduates with grossly inadequate training, loss of a full generation of teachers and scientists, and a profound weakening of the medical colleges at a time when scientific and clinical medicine were advancing extremely rapidly in the industrialized world.

The process of rebuilding the medical colleges began in 1975. The curriculum was extended to five or six years of medical studies; students were admitted primarily on the basis of academic performance after graduation from upper middle school; and the process of restoring academic quality began with the retraining of staff, the rebuilding of libraries, and the encouragement of research. By 1982, there were 116 medical colleges in operation enrolling approximately thirty thousand new students each year in major professional programs of Western medicine, traditional Chinese medicine, public health, pediatrics, stomatology, and pharmacy. Smaller numbers of higher-level personnel were also being trained in forensic medicine, nursing, medical technology, clinical nutrition, health economics, and health statistics. (The growth in the numbers of different types of health personnel and of training institutions during the period from 1949 to 1981 is illustrated in table 4.)

In the secondary medical schools, half of the students are currently enrolled in nursing programs and half in training programs for assistant doctors (in Western medicine, public health, maternal and child health, and traditional Chinese medicine), assistant stomatologists, assistant pharmacists, midwives, radiographers, and technicians. The students are usually junior middle school graduates, and their courses provide two to three years of academic work and a year of clinical internship, usually in county health or hospital facilities. Since 1980, assistant doctors who have graduated from secondary medical schools may be promoted to the standard of senior doctor by passing a qualifying examination after five or more years of experience and in-service training or part-time courses.

Table 4

Growth in Health Personnel and Educational Institutions, 1949–1981

Health Personnel	1949	1952	1965	1980	1981
(A) High Level					
Western doctors	38,000	51,736	188,661	447,288	516,498
Traditonal Chinese doctors	276,000	306,000	321,430	262,185	289,502
Integrated doctors[a]	--	--	--	--	1,591
Western pharmacists	484	900	8,265	25,241	29,948
Traditional pharmacists	--	6,536	71,848	106,963	132,805
Laboratory personnel[b]	--	--	--	--	20,987
Others	391	860	6,476	29,493	18,783
Total	314,875	366,032	596,680	871,170	1,010,114
(B) Middle Level					
Assistant doctors	49,400	66,500	252,713	443,761	436,196
Assistant pharmacists	2,873	7,071	37,201	83,901	81,863
Nurses	32,800	60,900	234,546	465,798	525,311
Midwives	13,900	22,400	45,639	70,843	70,904
Laboratory technicians[b]	--	--	--	--	63,242
Other	4,304	11,316	49,771	110,132	53,891
Total	103,277	168,187	619,870	1,174,435	1,231,407
(C) Primary Level[c]	86,888	156,218	315,045	752,636	769,517
Total Technical Staff	505,040	690,437	1,531,595	2,798,241	3,011,038
Educational Institutions					
Medical colleges[d]	56	31	92	112	116
Secondary medical schools	--	--	298	555	556

Source: The Health Sector in China.

[a] Graduates of both Western and traditional colleges.
[b] Laboratory staff included in "Others" in 1949, 1952, 1965, and 1980.
[c] Excluding "barefoot doctors" and part-time brigade workers.
[d] Excluding four medical colleges responsible to the army and three new provincial medical colleges that have not yet accepted students.

It has been estimated that up to 20 percent of assistant doctors became senior doctors by this route.

Administrative Responsibility for the Medical Colleges

Administrative responsibility for thirteen of the medical colleges rests directly with the Ministry of Public Health. Provincial or autonomous region health and education bureaus oversee ninety-six others; and seven are managed by specific enterprises, such as railways, metallurgy, and coal mining. Only four of the provincial programs are faculties of Western medicine within general universities; the remainder are independent medical colleges. Three new provincial medical colleges are under development. The training programs are from five to six years' duration in most medical colleges, but approximately 20 percent of students enroll each year in twenty-four medical colleges that have three-year courses. Most of these three-year colleges are secondary medical schools that were upgraded at the time of the Cultural Revolution. The Army also operates four medical colleges.

The Division of Medical Education and Science of the Ministry of Public Health designs manpower plans, prescribes curriculum, and prepares teaching materials for use by the medical colleges. As a result, there is a high degree of conformity in the content of educational programs. The Ministry of Education has responsibility for principles and policies relating to the educational programs of the medical colleges, for authorizing new specialties in undergraduate and graduate training, for assigning graduates, and for promotion of faculty to the rank of full professor. Recommendations are submitted by the Ministry of Public Health to the Ministry of Education for approval and incorporation into the higher education plan. Each provincial health bureau has a division of medical education and research which is responsible for programs in the medical colleges, hospitals, and research institutes of that province and which makes recommendations to the provincial bureaus of higher education and health. The education system is centrally controlled, resulting in nationwide conformity of programs.

The thirteen medical colleges administered directly by the Ministry of Public Health are among the strongest in the system. With the exception of the Capital Medical College (formerly, Peking Union Medical College), these "core" medical colleges have large undergraduate enrollments, but the ratio of professors and associate professors to students is nearly five times greater than that in the provincial medical colleges. The "core" medical colleges enroll one-half of all the postgraduate, master's, and doctoral candidates (the future teachers for the system), have the strongest research programs, and serve as national or regional training centers. Two of the thirteen are colleges of traditional Chinese medicine.

The "core" medical colleges receive their operating budgets from the Ministry of Public Health based on a formula (2,200 *yuan* per student in 1982) that is supplemented by appropriations to cover the operating costs of affiliated teaching hospitals and new capital construction. The remaining medical colleges are funded, at a significantly lower level, from the budgets of the provincial bureaus of education and health. All the medical colleges compete for research support from a Ministry of Public Health fund (8 million *yuan* in 1982), and the "core" medical colleges claim a disproportionate share of these resources. In addition, the ministry has financial responsibility for the Chinese Academy of Medical Sciences, which comprises twenty institutes and four hospitals in specialized subject areas of Western medicine, and for the Chinese Academy of Traditional Medicine, which has three institutes and two hospitals. In addition to carrying out basic and clinical research, the staff of both academies provides technical guidance to the medical colleges and affiliated hospitals.

Health Manpower

The Ministry of Public Health estimates that the ratio of senior doctors is approximately 0.8 per thousand population and that this ratio can be maintained during the 1980s without increasing medical college enrollments beyond the current level of approximately thirty thousand new admissions per year. Projections for the number of graduates from each of the medical college programs for the years 1982, 1985, and 1990 are listed in table 5. The priority for the current decade is consolidation rather than expansion, with emphasis on the quality of education and research, repairing the damage from the Cultural Revolution, and establishing a supply of future teaching and research staff through postgraduate training. In the ensuing decade, it is proposed to expand the educational programs in order to achieve a senior doctor to population

Table 5

Medical Colleges: Projections of Manpower Supply

Medical Colleges	Graduates Expected in 1982	Graduates Expected in 1985	Graduates Expected in 1990
Medicine	21,098	18,351	
Public health	1,183	1,345	Increase by 50%
Pharmacy	1,385	1,290	
Pediatrics	266	440	
Stomatology	462	677	Increase by 100%
Traditional medicine	4,388	3,944	
Traditional pharmacology	917	740	
Others	264	1,378	
TOTAL	29,963[a]	28,165	Increase by 10%

Source: The Health Sector in China.

[a] These numbers are unusually high due to temporarily expanded enrollment in 1977–78.

ratio of one per thousand by the year 2000. This target could be met by expanding medical college output by a further twelve thousand graduates per year (approximately 35 percent) or by working to upgrade large numbers of assistant-level doctors.

There are ninety-two faculties of Western medicine in the medical colleges and three new faculties under development. Projections by the Ministry of Public Health indicate a 13 percent decrease in enrollments by 1985, followed by a 10 percent increase by 1990. These colleges have a particularly heavy burden with respect to the retraining of the large number of doctors who graduated during the Cultural Revolution. In addition, they face the challenge of upgrading their staff to cope with the vast body of medical science that has emerged during the past twenty years. Approximately one hundred medical college staff with foreign-language capability are selected each year for training abroad.

Faculties of public health exist in twenty-four medical colleges, with a total enrollment of approximately fifteen hundred new students each year. It is proposed that the number of graduates be increased by 10 percent in 1985 and by 50 percent by 1990. This is intended to overcome the current shortage, which was accentuated by the loss of doctors from public health practice during the Cultural Revolution. The other source of public health doctors will be the upgrading of assistant doctors of public health who have graduated from secondary medical schools. In the manpower plans for public health, little attention has been given to the changing nature of public health practice caused by the shift in patterns of morbidity and mortality from acute infectious diseases to chronic diseases.

A separate undergraduate training path in pediatrics exists in ten medical colleges, but the total number of graduates is less than three hundred per year. No new faculties of pediatrics are proposed, but a modest increase in the number of graduates from existing faculties is projected by 1990.

There are thirty-three faculties of traditional Chinese medicine, twenty-four in independent colleges devoted exclusively to traditional medicine and nine in medical colleges that also offer training in Western medicine. The expected output of graduates—4,388 in 1982—will be insufficient to maintain the current number of senior doctors of traditional Chinese medicine in practice. Consideration is now being given both to the expansion of enrollment in existing faculties and to the establishment of new colleges of traditional Chinese medicine. These colleges acknowledge the need to strengthen substantially the scientific capability of the teaching staff and to promote research on the distinctive diagnostic and therapeutic techniques of traditional medicine. But traditional Chinese medicine is the one program of medical education where there is strong political pressure for rapid expansion of enrollments. This poses a serious problem in relation to the objective of improving the quality of teaching and research at these colleges. The colleges of traditional Chinese medicine also train traditional Chinese pharmacists, and some offer courses in the specialties of acupuncture and massage. In keeping with the principle of complementarity of Western and traditional Chinese medicine, medical students in the Western colleges receive approximately 140 hours of instruction in traditional Chinese medicine. One-third of the curriculum for students in colleges of traditional Chinese medicine concerns the basic sciences, clinical techniques, laboratory methods, and pharmacology of Western medicine. A small number of practitioners are fully trained in both Western and traditional Chinese medicine.

Stomatology is offered as a separate undergraduate program of six years' duration in five of the "core" medical colleges operated by the Ministry of Public Health and as a four- and five-year program in thirteen provincial medical colleges. Stomatologists are in short supply, and the Ministry of Public Health proposes to increase medical college output by 50 percent in 1985 and by 100 percent in 1990 in order to improve the ratio of stomatologists to population from 1:150,000 to 1:75,000. The stomatology programs also face a substantial challenge in upgrading the quality of both the educational experience and the training of teachers and scientists.

Four medical colleges offer full undergraduate programs in Western pharmacy, and there are, in addition, two independent schools of pharmacy. The Ministry of Public Health has not given priority to the expansion of enrollments in existing programs or to the addition of new faculties of pharmacy.

The projections for graduates from secondary medical schools are listed in table 6. The 1982 figures are misleading, since graduating classes in that year were unusually large due to an overly enthusiastic national policy in 1979 to expand enrollment. The 1985 figures reflect current policy. There is no plan to alter the number of secondary medical schools, but a substantial shift in enrollments is reflected in the projections, in particular, the increase in the number of nurses, assistant

Table 6

Secondary Medical Schools: Projection of Manpower Supply, 1982–1990

	Graduates Expected in 1982[a]	Graduates Expected in 1985[b]	Graduates Expected in 1990[c]
Assistant Western doctors	5,559	4,132	4,118
Assistant doctors of traditional Chinese medicine	6,464	1,760	1,830
Nurses	40,742	28,060	27,085
Assistant doctors of public health	3,085	2,187	2,720
Assistant dentists	428	931	1,130
Others	14,870	18,613	19,656
TOTAL	71,148	55,683	56,539

Source: The Health Sector in China.

[a] All figures for graduates in 1982 are considerably higher than normal due to large enrollment of students in 1979.
[b] Expected graduates for 1985 are lower than normal, reflecting provincial decisions on enrollment in 1982 influenced by the unusually large output of graduates in 1982.
[c] Figures for 1990 indicate expected outputs if current MOPH policies are implemented by the provinces.

doctors of public health, and assistant stomatologists and the reduction in the number of assistant doctors of Western medicine and traditional Chinese medicine. Since the ratio of senior-level doctors to population in rural areas is only one-tenth the ratio obtaining in urban centers, assistant doctors are a critical element in access to medical care for the rural population. The dispersed geographic distribution of the secondary medical schools has facilitated access to training for students from rural areas and increased the supply of graduates to these areas.

Challenges in Medical Education

The Learning Process

Considering the disruption caused by the Cultural Revolution and the limited financial and technical resources available, remarkable progress has been made since 1977 in the redevelopment of sound educational programs in medicine. Problems that were typical of medical schools in Western countries in the 1950s and 1960s, however, remain. The classes are large; all students follow the same course of study with very limited opportunities for independent learning or electives; the approach is didactic, with the dominant emphasis on transfer of knowledge and much more limited attention given to skills and attitudes; knowledge is organized by discipline, with considerable overlap between courses but inflexibility about substitution of new subjects; knowledge is drawn primarily from prescribed texts, with scant use of library materials or independent learning resources; and students and staff alike are strongly influenced by the exit requirement—a national examination of four hundred multiple-choice questions first introduced on a pilot scale in 1982. Major gaps in teaching have been recognized in molecular biology, genetics, immunology, virology, social and behavioral sciences, child development, geriatrics, and rehabilitation. Teaching in epidemiology, preventive medicine, and public health in the medical colleges is concerned primarily with infectious diseases and environmental pollution, although advanced epidemiological work is under way on cancer and on cerebrovascular and cardiovascular diseases in the academy research institutes. Faced with such large enrollments, staff who are unavoidably out of date, and lack of modern teaching and research equipment, it is a formidable challenge for the colleges to update the curriculum, move students from a passive role to active participation in the learning process, and introduce independent learning, problem-solving, and elective courses.

The Ministry of Public Health has concentrated its attention on the "core" medical colleges to launch the process of modernization. Most of the colleges have adopted sound educational objectives, but there is less confidence about translating these objectives into practice. There is considerable insitutional inertia due to the sheer size of classes and the enormous job of handling today's problems: the shortage of staff equipped to handle the new areas of knowledge; the slow turnover of academic staff due to life-long tenure; the impediment, until very recently, of political administrators with little sense of academic and professional objectives in charge of most institutions; and the obsolete equipment and facilities. Three important changes have occurred. First, younger professional leaders have been appointed to the position of president in many of the medical colleges. Second, a number of important educational innovations have been initiated in the "core" medical colleges. For example, Zhongshan Medical College has initiated two parallel streams of education in its undergraduate program: one group will follow the regular curriculum, and the other will have reduced lectures, more time for independent learning, and faculty tutors committed to this approach. Independent study is advocated by most of the "core" medical colleges, but less attention has been directed to the problem of motivating students to change learning habits. Regional production centers for audiovisual resources are planned at selected colleges. Recognition of the importance of studying the process of medical education and achieving better methods of setting objectives and evaluating results is reflected in the decision of the Ministry of Public Health to establish at China Medical University, Zhongshan Medical College, Shanghai First Medical College, and Beijing College of Traditional Chinese Medicine educational development units as a resource for both local and national needs.

A danger apparent to most observers is that medical colleges may be attaching too much importance to acquisition of sophisticated scientific and teaching equipment as a means of solving the current problems of medical education. The goal of having students in medicine and public health trained to operate sophisticated instruments that are unlikely to be available to them in practice is a dubious educational objective. The medical colleges seem to be underestimating the importance of setting educational objectives, of motivating both faculty and students to adopt changes in the approach to teaching and learning, and of ensuring that the evaluation of students and faculty reinforces the objectives.

National multiple-choice computerized examinations will be used to evaluate the students graduating from more than fifty medical colleges in 1984. In view of the profound steering effect of exit examinations on curriculum in medical colleges and on the attitude of students towards the teaching program, it is of overriding importance to evaluate the design of the examination to ensure that it reinforces, and does not negate, the educational objectives of the medical colleges and the Ministry of Public Health. At the moment, the ministry designs a content-oriented curriculum which is followed in most medical colleges and provides the principal teaching materials for that program. If the colleges adopt greater diversity of educational approach and introduce greater emphasis on professional skills and attitudes, it will be important to modify the examination process to reflect these objectives in the evaluation of graduating students.

Strengthening Scientific Capability

A second major challenge is strengthening scientific capability in the medical colleges and, in particular, rectifying the serious gaps in the new areas of science that have become so important in medical education and medical care during the past twenty years. In many of the "core" medical colleges, however, the opportunities for retraining and for acquisition of equipment are being used to strengthen traditional disciplines. The necessity of training teaching and research staff in basic medical science areas, such as molecular biology, immunology, and virology, is recognized. Yet there has been much less interest to date in other fields that are of critical importance to China, such as child development and social and behavioral sciences relevant to health. In preventive medicine and public health, established areas of interest in infectious diseases and environmental pollution still dominate educational programs, and there has been less adaptation of these important disciplines to noninfectious chronic diseases. Consequently, students, particularly in medicine, have difficulty in seeing the relevance of these educational programs to clinical practice. Finally, there has been very little attention given to evaluative epidemiology in the scientific assessment of disease burden, efficacy and effectiveness of preventive, diagnostic, and therapeutic interventions, and management of health resources.

The approach to scientific development is primarily "technology-push" rather than "problem-pull." The appetite for sophisticated scientific apparatus is enormous. Less attention is being given to the identification of key problems to be resolved by research and to concentration of resources in selected fields as a strategy to achieve high-quality research and training in fast-moving fields of scientific development.

The "core" medical colleges are faced with the traditional academic dilemma, the tension between science for its own sake and research focused on national or local problems. Whatever balance is struck, there is already evidence that local relevance may be an important route to international eminence. The epidemiological mapping of diseases such as cancer, heart disease, and hypertension has revealed geographic areas of high prevalence, and this may speed identification of causes and successful management. From the treasure house of traditional Chinese medicine have come drugs of international importance, such as *qinghao su* (extracted from wormwood, *Artemisia apiacea*) for malaria. Looking ahead, the application of evaluative epidemiology will be important in establishing the efficacy and effectiveness of Western and traditional Chinese medicines and in shaping a more efficient system of health services.

Preventive Strategies

A third challenge is to strengthen preventive strategies for noncommunicable diseases. Improvements in life expectancy, control of many common infectious diseases, and the one-child family policy call for the development of new approaches to the teaching of preventive medicine and health promotion in medicine, public health, and pediatrics. Although substantial work has been done in some of the institutes of the Chinese Academy of Medical Sciences, these specialties have not adapted their practices to use the results of this research in the prevention of such common, serious, noncommunicable diseases as hypertension, chronic obstructive pulmonary disease, cancer, and heart disease; to the protection of the child during antenatal and postnatal development; and to the prevention of illness in and rehabilitation of the growing population of elderly. Since the disease profile in China is rapidly coming to resemble the pattern in Western industrialized countries—but with only a fraction of the financial resources available to meet the needs—it is of critical importance to delineate those techniques of primary and secondary prevention which will delay the incidence and reduce the severity of these disorders. Since the emphasis in the medical colleges is on "technology-push," these problems are not being given the attention they deserve in training and in research.

Effectiveness and Efficiency

A fourth challenge is the development of health services evaluation and management capability. In both the developed and the developing world, the capability to evaluate scientifically the effectiveness of new and traditional techniques of prevention, diagnosis, and treatment is assuming much greater importance because of their uncertain impact on health and the high cost of medical technology. The "core" medical colleges, which set the pattern for the system, have a special responsibility to develop this evaluation capability since new medical techniques are likely to be implemented first in these centers. This type of evaluative research is relatively unfamiliar to the medical colleges. Furthermore, there has been some resistance to randomized, controlled trials and political sensitivity about evaluation of the practices of traditional Chinese medicine. Nevertheless, it is a high priority for colleges of Western medicine and traditional Chinese medicine to develop evaluation capability closely linked to clinical programs, where it is needed both for improvements in the effectiveness of practice and in the teaching of undergraduate and postgraduate students. To enhance the development of this capability, the Ministry of Public Health has authorized overseas training and technical assistance for staff at each of the "core" medical colleges and established three national training centers in design, measurement, and evaluation at Sichuan Medical College, Shanghai First Medical College, and Guangzhou College of Traditional Chinese Medicine. A high-level national seminar on this subject was held in April 1984.

Scientific principles of management are at a very early stage of development in the Chinese health-care system. The medical colleges have the responsibility for training health administrators but lack individuals with a strong background in the management sciences. Because they are devoted exclusively to medical training, these schools also lack the interaction that can occur in multipurpose universities with groups in management studies, economics, and other relevant disciplines. Furthermore, the medical colleges have not been asked to address the key issues of management of the health-care system. Emphasis has been primarily on their own institutional administration.

The Ministry of Public Health has authorized the establishment of regional centers for management training, five in "core" medical colleges and one for the northeast at Harbin Medical College. The challenge at this stage will be to establish capable teaching staff through training abroad and technical assistance. Since management is an applied science, it will be important to develop case studies of the Chinese healthcare system and institutions in order to provide effective teaching materials for training. Furthermore, there is a need to stimulate healthcare research and operations research on management problems to provide evidence both for training and for improving on-site management.

Specialized Manpower Training

A fifth challenge is the implementation of a strategy for graduate-degree education of teachers and scientists and postgraduate clinical specialists that can achieve high standards and meet national manpower needs without diverting attention and resources from the reform of undergraduate programs.

An early priority in the upgrading of medical education is to improve the qualifications of existing teaching and research staff in the medical colleges and to train new staff. In 1982, more than two thousand students were enrolled in master's programs and a small number in doctoral programs at the medical colleges and academy research institutes. The Ministry of Public Health determines the eligibility of medical colleges to train graduate students in each discipline on the basis of the quality of the teaching staff and research achievements in that discipline at the college. The ministry also determines the number of students enrolled in each program. To date, these decisions have been based more on the strength of existing programs than on the need for personnel. As a consequence, relatively little training is being undertaken in the new subject areas in which the greatest demand for both teachers and research scientists exists. Moreover, most of the graduate degree programs have small enrollments and limited resources for research. It is unlikely that these programs will achieve the high standards that might be reached if resources and enrollment for selected disciplines were concentrated at a few colleges.

Similarly, the Ministry of Public Health has designated two hundred sites for advanced training in the clinical specialties in Western medicine. The ministry determines the number of trainees and the hospitals from which they will come, monitors the quality of the programs, and grants a certificate on completion of the three-year training period. The ministry also provides operating support to the affiliated hospitals in the form of a grant of 750 *yuan* per year for postgraduate trainees, together with capital investment in facilities, equipment, and library resources for the training sites. Half of the funds provided for support of trainees is actually provided by the hospital that sends the individual, except in the case of minority nationalities, where the full amount is paid by the ministry. Candidates for specialty training must have at least three years of clinical experience after graduation from medical college. On completion of training, the individual returns to the hospital from which he or she was nominated. At this time, there are no specialty examinations, but the possibility of certification is under consideration. Once again, the pressure for clinical specialty training is in the established areas of strength. The Ministry of Public Health has not yet established a manpower plan that provides guidelines for the numbers needed in various specialties, nor has there been an attempt to establish on a regional basis special incentives for training programs in underrepresented specialties.

In several leading "core" medical colleges, programs are planned or under way in nearly every discipline of science and every specialty of practice. Since these programs are closest to the hearts of faculty members, they will consume much of their attention at the risk of shortchanging undergraduate teaching. Furthermore, even the strongest medical college is unlikely to achieve high standards in a large number of programs. A high priority, therefore, is the rationalization of programs of graduate study and specialty training at this early stage of their development.

Continuing Education

Continuing education is one of the most difficult challenges facing the medical colleges. For example, in the field of management alone, it is estimated that 310,000 people are engaged in management and administration of health services and that almost all will require substantial upgrading of their skills. In the professional categories of medical manpower, it is essential to reach the doctors already in practice, in view of the rapidly changing pattern of health needs and the inadequacy of their original training to cope with the new challenges. To date, most continuing education has taken the form of courses at the medical colleges. In the period ahead, it will be imperative to reach out and assess the need for continuing education in communities, to provide courses and learning resources on-site, where individuals practice, and to provide some evaluation of the effectiveness of the various approaches to continuing education in addressing these needs.

A formidable but temporary problem in continuing education is the need to retrain more than 160,000 doctors who graduated from medical colleges during the Cultural Revolution. Approximately one-third of those undergoing retraining take full-time courses over a period of one and a half to two years to rectify deficiencies in premedical and basic science training. The remainder take part-time instruction over four or five years in medical school courses and through television education while they continue their service work in hospitals, epidemic prevention stations, research institutes, and government bureaus. Candidates in both the full-time and part-time retraining programs are given the title of visiting doctor after passing the final examinations. They do not receive the bachelor of medicine degree, which is reserved for regular graduates of the five-year medical colleges.

The Ministry of Public Health has initiated a series of medical education reforms supported by a World Bank health project. The areas singled out for attention are the quality and effectiveness of the educational process for initial professional training and for continuing medical education and the strengthening of teaching capability in preventive strategies, evaluation of effectiveness, and efficiency and management.

Conclusion

Chinese achievements in health since 1949 are remarkable by any standard, but they are particularly amazing considering the magnitude of the task and the size of the population served. Most of the problems that confront medical colleges in China are similar to those experienced in industrialized countries: coping with the phenomenal growth in knowledge relevant to medicine without overwhelming students with memorization of facts (returning to the "stuffed Peking duck" syndrome); introducing new pedagogic approaches to achieve more active student participation in the learning process; training a new generation of faculty; carving out niches for preventive medicine, social and behavioral sciences, and evaluative epidemiology in a curriculum dominated by biomedical science and therapeutic medicine; and rationalizing programs of graduate degree studies and postgraduate specialty training to achieve both the highest possible quality through concentration of resources in a smaller number of programs in each discipline and a reasonable balance of manpower production in relation to the needs of the medical colleges and health system. Yet China also faces unique problems: the relationship of parallel systems of Western and traditional Chinese medicine, and the technological upgrading of traditional medicine, the foundation of which is not explained by the laws of the natural sciences.

With the Four Modernizations, China has adopted an outward-looking posture for upgrading its medical colleges that is in keeping with the stated policy of self-reliance through international cooperation. Training abroad for large numbers of staff, foreign technical collaboration, and importation of state-of-the-art technology are all part of the effort to close the large technological gap as rapidly as possible. At the same time, substantial capital investments have been made in new facilities and equipment, and more dynamic leadership of the medical colleges is anticipated as the national policy of replacing political administrators with professionally qualified leaders is implemented. The initial investment has been heavily concentrated in the thirteen "core" medical colleges directed by the Ministry of Public Health. The benefits, however, must be measured by the ripple effect on the quality of education in the provincial medical colleges, which train more than 80 percent of medical personnel, and on the effectiveness, efficiency, and orientation of the system of health services. Mechanisms to achieve this broader impact must rank at the top of the agenda for future action.

A

Academia Sinica. *See* Chinese Academy of Sciences
Academies of agricultural sciences, provincial, 197
Academy of Agricultural Sciences, applied research, 197
Acta Sinica divisions, 197
Acupuncture and massage, 249
Agricultural Research Institute, pest protection labor, 188
Agriculture: animal manure, 177–78; aquaculture, 177; arable land, 174; broad spectrum insecticides, 192; capital costs of mechanization, 184; cereal production, 185; chemical fertilizers, 178; communes and human waste, 177; crop domestication, 180; crop improvement, 180–81; crop selection, 180; ecological locust control, 191; Eight-Point Charter, 196; fertilizers, 177–78; food crisis, 209; food production, 184–85; four-level network, 198, 200; Four Modernizations, 196, 198–200; geography and climatology, 172, 174, 176; government control, 203; grain production, 216; "hydraulic," 180; international trade, 208; irrigation, 176–77; land reform and inequality, 216, 218; locust control, 190–92; manures, *Tab.*, 179; mechanization, 176, 183–84; new introductions, 172; noncereal statistics, 185; pest protection labor, 188; plant breeding, 180–83; plant pest forecasting, 188, 195; principal crops, 172–74; protection from pests, 185, 188, 190; research and education, 195–200; research and technology, *Fig.*, 199; research organization, 196–200; rice paddy ecology, 193, 194; rice pest management, 192–95; rice pest management success, 194; rice pest catastrophes, 193–94; university and college research, 200; water control, 172; work force, 176
American Committee for China Famine Relief, Nanking–Cornell program support, 24
American Society for Human Genetics, 161
Amoy, University of, herbarium, 45
Andersson, J. G.: Geological Society of China, 72; mining advisor to Geological Survey, 68–69; vertebrate paleontology, 81
Anhui Province, schistosomiasis study, 133
Anthropology, Choukoutien, 82

Anthropometry, 218–227
Antischistosomiasis campaign, 130
Apple, 173
Apricot, 173
Archaeology, Choukoutien, 82
Armyworm, 192
Arnold Arboretum, Harvard University, 43, 51–62; exchange relations, 51; specimen exchanges, 58–60, Tab., 59; education, 32
"Aspiration Society" (Shang-chih Hsüeh-hui), Fan Memorial Institute of Biology funding, 50
Atherosclerosis, 184

B

Babbitt, Irving, Harvard University, 56
Baltimore, reverse transcriptase, 164
Beadle, George, Cornell University geneticist, 24, 159
Bean, velvet, 173
Beijing Biological Production Institute, vaccines, 153
Beijing College of Traditional Chinese Medicine, educational development, 252
Benzene hexachloride. See BHC insecticide
BHC insecticide, 191, 192
Biography, dictionary project, xviii

Biological Laboratory of the Science Society of China, Nanking, herbarium, 44
Biological Supply Service: T'an Chia-chen's exhibits, 21; Wu Chin-fu as head, 17
Biologists, financial support for, 41
Biology, methodology, 8, 13–14
Black, Davidson, Choukoutien, 82
Borer (rice pest), 192
Boring, Alice, biology department head, Yenching University, 15; taxonomy, 9, 18, 41
Boring, Edwin, Harvard University, 41
Botanical Congress, Fifth International (1930), 31
Botanical institutions, funding patterns, 46–51
Botanical materials: competition, 62; exchange, 34, 35–38; exchange relations, 39, 51, 62
Botanical Museum of Berlin, exchange relations, 51
Botany, 92; local, 92; taxonomic specimens for study, 38; taxonomy, 32, 33–35, 40; Western research in China, 40
Botany funding: China Foundation for the Promotion of Education and Culture, 46; foreign research institutions, 46; philanthropists, 46;

Botany funding

research centers, 42–45;
Rockefeller Foundation, 46
Boxer Indemnity, 39, 49
Bretschneider, Emil, seed collecting in China, 52
Bridges, Calvin B., Columbia University geneticist, 156, 158; and Li Ju-ch'i, 19–20
Boxer Indemnity fund, 24–25
British Museum of Natural History, systematic photography, 36
Britton, Nathaniel Lord, 55
Buck, J. L., Szechwan Agriculture Improvement Institute, 26
Buck, Peter, xv–xvi; Chinese conditions, 93
Burbank, Luther, 160; plant selection, 161
Burnham, Charles, geneticist, 159

C

CAAS. *See* Academy of Agricultural Sciences
California, University of, Berkeley, 43; exchange relations, 51
Caloric intake, *Tab.*, 186
Capital Hospital. *See* Capital Medical College of China
Capital Medical College of China, 245; gossypol study, 116
Carbamate insecticides, 193

Ch'en Huan-yung

Carbofuran, 194
CAS. *See* Chinese Academy of Sciences
Case, Marion, Case Estates of the Arnold Arboretum, 53
Chang Hsi-chih, invertebrate paleontologist, 81
Chang Hung-chao, 67; geology, 68; paleontology, 83
Chang Keng, Research Institute of Geology of Academia Sinica, 75
Chao, Edward C. T., U.S. Geological Survey, 66
Chao Lien-fang, Szechwan Agriculture Improvement Institute, 26
Ch'en, Jerome, on Chinese interest in synthesis, 91
Ch'en Chen: experimental approach, 9; goldfish, 9–14; National Southeastern University, first genetics course in China, 10; teaching role, 12; Tsinghua University biology department head, 11–12
Cheng, T. H., 131
Cheng Hou-huai, Chungyang University, 75–76
Ch'en Huan-yung, 33; and the Arnold Arboretum, 53; and Western botany, 40; botanical taxonomist, 39; Harvard loan repayment, 55–56; Harvard Sheldon Traveling Fellowship, 54; Harvard University Bussey Institution for Research in

Applied Biology, 53–54; history of botany, 31; National Sun Yat-sen University Botanical Institute, Sino-Japanese War, 44, 63; need for support, 42; plant collector for Arnold Arboretum, 54
Chen Shu, invertebrate paleontologist, 80
Ch'en Tze-ying: Morgan influence, 159; National Amoy University biology department head, 16–17; research, 16; Yenching University biology department, 15, 16–17
Chen Ziying. *See* Ch'en Tze-ying
Chiang Mo-Lin, Peking University, 68
Chidanken, 79
Ch'ien Sung-shu: and Western botany, 40; botanical taxonomist, 39; National Central University herbarium, 44; need for support, 42
Chigaku dantai kenkyūkai. *See* Chidanken
Children, heights and weights, *Tab.*, 221
China: bulk steroids, 111; cradle of agriculture, 171; development indicators, *Tab.*, 204–5; grain trade, *Tab.*, 211; history, interest in, 91; life expectancy, 201–2; nutrient availability, *Annex Tabs.*, 230–31; rice bowl, 194; self-sufficient in food production, 184; synthesis, interest in, 91; vegetarian emphasis, 184
China Foundation for the Promotion of Education and Culture, 4; botanical institutions, 46; botanists' support, 39; Ch'en Tze-ying support, 16–17; Fan Memorial Institute of Biology funding, 50; funds for studying flora and fauna, 49; graduate student support, 18; Li Ju-ch'i support, 20; Nanking-Cornell program support, 24; National Southeastern University, 11; naturalistic/localistic bias, 7; Sino-Japanese War, 63; taxonomy funding, 51; Tsinghua University, 11
China Foundation Science Research Professor Program: Ch'en Huan-yung at National Sun Yat-sen University, 50; Hu Hsien-su at Fan Memorial Institute, 50–51
China Medical Board, 4, 5; science education plan, 18; taxonomy, 7; Tsinghua University, 12; Yenching University, 14–15
China Medical University, educational development, 252
Chinese: cabbage, 173; chestnut, 173; hickory nut, 173

Chinese Academy of Medical
  Sciences, 246; Institute of
  Parasitic Diseases, Shanghai,
  128; preventive medicine, 254;
  reproductive health, 100–101
Chinese Academy of Sciences
  (formerly National Academy
  of Peiping), 70, 198; Institute
  of Developmental Biology,
  101; Institute of Geology, 68,
  71; local bias, early, 7; locust
  control, 191; reproductive
  health, 100–101; research
  organization, 196
Chinese Academy of
  Traditional Medicine, 246
Chinese government: Ministry
  of Agriculture and Forestry
  cooperative research projects,
  26; rural agricultural policy,
  24–25
Ching, R. C. *See* Ch'in
  Jen-chang
Ch'in Jen-chang: systematic
  photography, 39; Western
  botany, 40
Chin-ling nu-tzu wen-li
  hsüeh-yuan. *See* Ginling
  College
Chin Shao-chi, Geological
  Survey funds, 69
Chou Ch'eng-yao. *See* Zhou
  Chengyao
Chou En-lai. *See* Zhou Enlai
Choukoutien, vertebrate fossils,
  81–82
Christensen, Carl, Danish
  botanist, 39
Christian missionaries, 94
Ch'ung-ch'ing ta-hsueh. *See*
  Chungking University
Chung Hsin-hsüan: botanical
  taxonomist, 39; need for
  support, 42; University of
  Amoy herbarium, 45
Chung-i yen-chiu-yuan. *See*
  Chinese Academy of
  Traditional Medicine
Chungking University, geology
  department, 68
Chung-kuo Chung-i
  yen-chiu-yuan. *See* Chinese
  Academy of Traditional
  Medicine
Chung-kuo k'o-hsueh-yuan. *See*
  Chinese Academy of Sciences
Chung-shan i-hsueh yuan. *See*
  Zhongshan Medical College
Chung-shan University
  (formerly Kwangtung
  University, National Sun
  Yat-sen University): botanical
  research, 43; geology
  department, 68
Chung Shao-hua. *See* Zhong
  Shaohua
Chung-yang University
  (formerly Tungnan), geology
  department, 68
Cinnamon, 173
CMB. *See* China Medical
  Board
CMCC. *See* Capital Medical
  College of China
Colleges. *See* Universities
Colon cancer, 184

Columbia University: genetics, 5, 158, 159
Communists, "rural reform" in Kiangsi, 25
Continuing education: doctors, 257–58; managers, 257; retraining, 258
Contraceptive pill for men: gossypol research, 114–18; life table analysis, 118
Contraceptives: availability, 102; improved technology, 103, 105; research on, 100, 105–118; prevalence, *Fig.*, 104; quantitative analysis of use, 102–3; reduced abortions, 105; use, 101–4
Cornell University: broad genetic training, 8; exchange program with University of Nanking, 5, 23–24; genetics, 5, 159; maize genetics, 158
Coulter, John M.: Chicago University naturalist, 7; Soochow science consultant, 18
Council of Graduate Schools, 237
Cramer, H. H., crop loss to pests, 185, 188
Crops: Central China, winter wheat, 174; fertilization, 172; Manchuria, 172; North China, 172, 174; domesticated, *Tab.*, 173; Sichuan, 174
Cucumber, 173
Cultivated land and population, *Fig.*, 175

"Cultural imperialism" in genetics, 6
Cultural Revolution, 57, 158, 185; disruption of medical education, 242–43; farm mechanization, 183; genetics, 156; plant breeding disruption, 181

D

Dana, James, *System of Mineralogy*, 66
DDT, 192
Death, causes, *Tab.*, 241
Degrees, bachelor's, science and engineering, 234
Dizhi Yanjiusuo. *See* Geological Institute
Dobzhansky, Theodosius, California Institute of Technology: genetics, 157; genetics synthesis, 8; and T'an Chia-chen, 21
Doctors' assistants, Western medicine, 243
Domestication of major crop varieties, 180
Drosophila pseudoobscura, 156
Dubinin, N. P., Mendelian geneticist, 162
Dyson, J. Y., Soochow University biology department, 17

E

East, E. M., Harvard University maize geneticist, 159
Economic geology, 72
Education: enrollment ratios, 233–34; foreign support, 234–37, government funds for construction, 234. *See also* Medical education
Emerson, R. A., Cornell University geneticist, 23–24, 159
Enders, J. F.: measles virus, 144; polio tissue cultures, 142
Encephalitis. *See* Japanese encephalitis
Engle, Hua-ling Nieh, translator, 124
Engle, Paul, translator, 124
EPI. *See* Expanded Program of Immunization
Ernst, Joseph W., xix
Expanded Program of Immunization, smallpox, 141
Experimental biology, 8, 92

F

Family planning research, 100
Fan Chien-chung, University of Nanking, historical approach to genetics, 25–26
Fang Yi, developmental biology, 101
Fan Memorial Institute of Biology, Peking, 44; and Sino-Japanese War, 62–63; funding, 50; herbarium, 45; specimens sent to Arnold Arboretum, 60; photography, 40
Fan Memorial Laboratory, taxonomy, 7
Faust, E. C.: schistosomiasis, 130, 137; schistosomiasis distribution, 127, 128
Feng Che-fang, Szechwan Agriculture Improvement Institute, 26–27
Fernald, M. L., Harvard University botanist, 36
Fertility rate, 100
Fertilizers: chemical, *Tab.*, 179; nitrogen recycling, 177
Field Museum of Natural History, Chicago, 40
Fisher, R. A.: genetics field designs, 161; population genetics, 159
Food energy requirements, *Tab.*, 211
Food policy, 203–9; and health, 203; availability, 210–14; consumption mix, 207–8; direct consumption of grains, 207; distribution, 214, 216, 218; government control, 203; grain transfer, 206; international agricultural trade, 208–9; issues, 227–29; rationing, 206–7; restricted

production of preferred foods, 207–8
Food production, *Tab.*, 187
Forecasting stations, locust migrations, 191
Four-Level Agroscientific Network, agricultural research results, 198
Four modernizations, 95, 196, 198–200
Fryer, John, translator, 66
Fukien Christian University, botanical research, 48
Furth, Charlotte, 65; historical science, 78
Fu Ssu-nien, naturalistic, 7

G

Gang of Four, 57
Garlic, 173
Gee, N. Gist, 33, 54; Biological Supply Service founder, 17; China Medical Board, 5; funding advice, 49–50; published with Boring, 18; Rockefeller Foundation, 48; Rockefeller Foundaton advisor, 41; and taxonomy, 7
Geneticists, 92; American Committee for China Famine Relief support, 41; Rockefeller Foundation support, 41
Genetic research, 4; fruit flies, 21; networks, 5–6; peasants, 164; T'an Chia-chen, 22
Genetics: American influence, 158–59; and food supply, 166; "applied" potential, 5; catching up, 168; coexistence, 162; Cultural Revolution, 163; education, 4; education networks, 5–6; frustrations, 167; genetic engineering, 167; "Johns Hopkins model," 5; Mendelian, 162, 163, 165; Michurin, 160–62; molecular biology, 162, 167; plant breeding, 167, 168; role of foundations, 4–5; Russian, 156; T'an Chia-chen, 20–23; taxonomy, 6–9; transposable genes, 165; Yenching University, 14–20
Genetics Institute, Fudan University in Shanghai, genetics, 156
Geological Institute, first school of geology, 67
Geological research, 66, 71–84; government support, 70
Geological Survey of China, 65; and China Foundation for the Promotion of Education and Culture, 69; Choukoutien, 82; division of Ministry of Industry, 70; geological mapping, 69; mineral resources, 69, 73; research, 68; Sinyuan Fuels Research Laboratory, 76
Geology, 91; applied, 72; education and institutions, 66–72; field science, 88–89;

Geology

local, 92; mineral resources, 67, 73, 74; prominence in China, 87–89
Geotectonics, 85–87
Ginger, 173
Ginling College, Nanking, herbarium, 45
Ginseng, 173
Goldfish, experimental animal, 13
Goldschmidt, Richard, 157
Gossypol: antifertility drug, 114; clinical trials, 115, 116; effects reversible, 115; oral use, 116; safety, 116; study, *Tabs.*, 117, 119, 120
Gould, Stephen J., 8
Grabau, Amadeus W.: "eustatic control," 83; Geological Society of China, 72; Geological Survey of China paleontologist, 80; geosynclines, 82; historical geology, 78, 83; pangaea, 83; Peking University geology department, 67
Grain, production increase, *Fig.*, 182
Grain-poor provinces, *Tab.*, 217
"Grass-roots geologists," 79
Gray, Asa: botanical theory, 52; Harvard University, 36
Gray Herbarium, Harvard University, exchange relations, 51
Great Leap Forward, 185; agricultural production, 213; farm mechanization confusion,

Health status indicators

183; fertilizer production, 178; medical schools, 242; schistosomiasis study, 127–28
Greene, Roger S., 49–50
"Green revolution," 181; plant breeding, 180; synthetic fertilizers, 178
Guangzhou College of Traditional Chinese Medicine, evaluation training, 255
Gui-chi County: Antischistosomiasis Hospital, 137; schistosomiasis, 133

H

Hainan Island, 44
Haldane, J. B. S., population genetics, 159
Harbin Medical College, management training, 256
Harlan, J. R., plant breeding, 166
Health, maternal and child, 243
Health manpower, 246–51
Health personnel: and educational institutions, *Tab.*, 244; training upgrade and expansion, 242
Health services: evaluation and management, 255
Health statistics, Shanghai County, *Tab.*, 241
Health status indicators, *Tab.*, 240

Henan Province: agricultural
  research, 198; Plant
  Protection Station, 195
Henan Teacher's College, plant
  pest management, 195
Hepatosplenic schistosomiasis,
  133
Herbaria: insect control, 35;
  type specimens, 36;
  photography, 36, 40
High blood pressure, 184
Higher education: enrollment
  increases, 235, 237; foreign
  experts, 235; government
  funds, 236; graduate degrees,
  237; quality, 235; research in,
  237; science and engineering
  improvement in, 237;
  student-faculty ratios, 235,
  237
High-yield strains of rice, 178
Hillcrest, Weston, MA,
  experimental farm, 53
Historical geology, 78–80;
  classification and induction,
  84; research methods, 78
Historiography, xv–xvi
Honan provincial geological
  survey, 71
Hong Kong Botanical Garden
  herbarium, 44
Horticulture: Chinese
  specimens, 37; commercial
  collecting, 37
Hsieh Chia-jung: mineral
  deposits, 76, 77; Sinyuan
  Fuels Research Laboratory,
  coal geology, 76

Hsi-pu K'o-hsüeh-yüan. *See*
  Western Academy of Sciences
Hsüeh-heng, journal, 56
Hua Heng-fang, translated
  Western geological work, 66
Huang Chi-ch'ing, invertebrate
  paleontologist, 80–81
Hu Han, Institute of Genetics
  in Beijing director, 164
Hu Hsien-su, 33; and Western
  botany, 40; Arnold
  Arboretum, 55; Arnold
  Arboretum research funds,
  58; botanical taxonomist, 39;
  cultural conservative, 56; Fan
  Memorial Institute of Biology
  exchange relations, 62; Fan
  Memorial Institute of Biology
  head of botany, 44; Harvard
  University Bussey Institution
  for Research in Applied
  Biology, 55; National Central
  University herbarium, 44;
  National Southeastern
  University biology
  department, 10; naturalist, 7;
  need for support, 42, 55;
  opposed to literary reform 57;
  Sino-Japanese War, 63;
  specimens sent to Arnold
  Arboretum, 60
Hunan provincial geological
  survey, 71
Hydraulic civilization, 177

I

Ijiri Shōji, paleontologist: and geology, 84; Chidanken, 79
Immunization. *See* Vaccination, Vaccines
Imperial University of Peking, 66
Infant mortality and food energy, *Fig.*, 215
Insecticide: application costs, 194; economic thresholds, 194; production, 188
Institute for Control of Pharmaceutical and Biological Products, JE vaccine, 147
Institute for Parasitic Diseases, Shanghai, schistosomiasis distribution, 132
Institute for Plant Protection, Kungchuling, 197
Institute of Biology of Academia Sinica, Nanking, herbarium, 44
Institute of Botany, Peking, herbarium, 45
Institute of Genetics, Chinese Academy of Sciences, 164–66; molecular genetics, 165; organization and research, 165; plant somatic cell genetics, 165; reports, 165–66
Institute of Geomechanics, Chinese Academy of Geological Sciences, 87
International Medical School in Guangzhou, Western medicine, 242
International Rice Research Institute, 181; germ plasm resources, 181
Intrauterine devices: acceptors, *Tab.*, 108; comparisons, 107; Copper T, 107, 109; improvement, 105–9; research, 105–9; termination, *Tab.*, 110
Invertebrate paleontology, 81
IRRI. *See* International Rice Research Institute
Irrigation, 172
Ishihama, Ota ring, 105–7
IUD. *See* Intrauterine devices

J

Jack, John: Arnold Arboretum and Bussey Institution, 54; support for collecting, 60; botanical lecturer, 53; Harvard University dendrologist, 33; and Hu Hsien-su, 55
Japanese encephalitis: inactivated vaccine, 147; Japanese vaccine, 147; live-attenuated vaccine, 148–49; vaccines, 146–49, *Fig.*, 150
Japanese Society of Parasitology, schistosomiasis control, 128
JE. *See* Japanese encephalitis
Jen Hung-chün, local geological materials, 88, 92
Jews, 48

Jilin Academy of Agricultural and Forestry Sciences, 197
Jilin Province, agricultural research, 198
Juan Wei-chou, coal deposits, 74

K

K'ai-lan, mining company, Geological Survey funds, 69
Kelman, Arthur, xix
Kerckhoff Laboratories, and T'an Chia-chen, 21
Kevles, Daniel, historian, 89
Kiangsi, provincial geological survey, 71
King, Sotsu G. *See* Chin Shao-chi
KMT. *See* Kuomintang
K'o-hsüeh nung-yeh, 26
K'o-hsüeh (Science), 57
K'o-hsüeh yü jen-sheng-kuan, xvii
Komiya, Y., schistosomiasis control, 128
Kuang-chou Chung-i hsueh-yuan. *See* Guangzhou College of Traditional Chinese Medicine
Kumquat, 173
Kuo-li chung-yang ta-hsueh. *See* National Central University
Kuomintang, and research institutions, 70
Kwangsi, University of, Research Institute of Biology, 44
Kwangsi provincial geological survey, 71
Kwangtung provincial geological survey, 71

L

Lampton, David, schistosomiasis politics, 139
Leafhopper, 193
Lee, J. S. *See* Li Ssu-kuang
Li, Choh-ming, xix
Li Chung-chün. *See* Li Zhongjun
Li Hsien-wen: Szechwan Agriculture Improvement Institute, 26, 27; University of Nanking plant geneticist, 25
Li Ju-ch'i (Li Ruqi): Beijing University geneticist, 159; *Drosophila* genetics, 156; ladybird beetles, 20–21; on Lysenkoism, 29; Yenching University geneticist, 18–19
Lindgren, Waldemar: Massachusetts Institute of Technology economic geologist, 75; mineral deposits, 76
Ling Li, Szechwan Agriculture Improvement Institute, 27
Lingnan University (formerly Canton Christian College), botanical research, 43, 48

Lin Hsing-kuei, and Geological Survey funds, 69
Lin Piao, 57
Li Ruqi. *See* Li Ju-ch'i
Li Shih-tseng, scientific research, 70
Li Ssu-kuang: geophysics, 86; invertebrate paleontologist, 80; Peking University geology department, 67; Research Institute of Geology of Academia Sinica, geophysical approach, 71; structural geology, 84; tectonic study, 85
Litchi, 173
Li Zhongjun: Beijing University population geneticist, 159, 161; University of Pittsburgh, 161
"Local sciences," 92
Locust, migratory, 190–92
Logan, O. T., schistosomiasis, 125, 127
Lotus, 173
Love, H. H., University of Nanking–Cornell University program, 24
Lycosa spiders, 192
Lyell, Charles, *Elements of Geology*, 66
Lysenko, T. D., 160
Lysenkoism, 28, 155, 157; appeal, 160; crosses, wide, 162, 165; statistical methods, 161

M

Ma Yinchu, demographer, 99
McClintock, Barbara, 8; Cornell University geneticist, 24, 159
McClure, Floyd Alonzo: Arnold Arboretum, 58; botanist, 42; Lingnan University herbarium, 44; need for support, 42; U.S. Department of Agriculture plant explorer, 42
MacGowan, John, translator, 66
Maize, 180
Malnutrition, 213–14, 223, *Fig.*, 225; food and disease, 226
Malthusianism, 99
Mao Shoupai, Institute of Parasitic Diseases, schistosomiasis control, 130
Mao Shou-p'ai. *See* Mao Shoupai
Mao Tse-tung. *See* Mao Zedong
Mao Zedong, 162; farm mechanization, 183; on schistosomiasis, 123, 124
Marx, Karl, xv
Manpower, medical, 249–251, 256–257
Mayr, Ernst, Harvard University geneticist, 3, 48
Measles, 144–46; vaccine technology, 145–46
Meat, consumption, 229
Medical care, rural assistant doctors, 251

Medical colleges:
  administration, 245–46;
  manpower supply, *Tab.*, 247;
  rebuilding, 243; research, 254
Medical education: child
  development, 253; chronic
  disease, 253; continuing,
  257–58; effectiveness and
  efficiency, 255–56;
  epidemiology, 253; evaluation,
  253, 255; health services
  administrators, 255; pedagogy,
  251–53; preventive medicine,
  253, 254–55; problem solving,
  252; scientific capability, 253;
  social and behavioral sciences
  in, 253; specialized, 256–57
Medical schools, secondary,
  243; graduate projections,
  249; manpower supply, *Tab.*,
  250
Medical techniques, evaluation,
  255
Medicine, preventive, 242, 254
Medicine, Western: and
  traditional medicine, 242;
  faculties, 248; services limited,
  242
Mei Kuang-ti, cultural
  conservative, 56
Meleney, H. E., schistosomiasis,
  127, 130, 137
Mendelism, 162, 169
Meng Hsien-min, Research
  Institute of Geology,
  Academia Sinica, 75
Merrill, Elmer Drew, 32, 42;
  Fukien Christian University,
  specimen identification, 45;
  Harvard University, herbaria
  insect control, 35; Lingnan
  University herbarium, 43;
  New York Botanical Garden
  specimens policy, 60;
  University of Amoy, specimen
  identification, 45; University
  of Nanking herbarium, 45
Metcalf, Franklin Post: Arnold
  Arboretum, 58; Fukien
  Christian University botanist,
  42; Fukien University
  herbarium, 44; Lingnan
  University herbarium, 44;
  need for support, 42
Methyl parathion, 193, 194
Metropolitan Museum of
  Natural History, Academia
  Sinica, Nanking, 45
Michurin, I. V., Russian
  horticulturist, 28, 160
Microbial insecticides: *Bacillus
  thuringiensis*, 188; *Beauvaria
  bassiana*, 188
Midwives, 243
Millet (*Setaria italicum*), 173;
  first domesticated food plant,
  171
Miner, Luella, Geological
  Society of China, 72
Ministry of Agriculture, 197;
  education and research, 235;
  locust control, 191; pesticide
  evaluation, 188; pesticide
  training programs, 188; radio
  programs for farmers, 236;

Ministry of Agriculture
secondary schools, 236; T'an Chia-chen, support for, 22
Ministry of Communication, 73
Ministry of Education: agricultural research, 236; medical college principles and policies, 245; medical education, 236; polytechnical colleges funds, 235; radio and television universities, 235
Ministry of Industry, Bureau of Mines, first school of geology, 67
Ministry of Public Health: clinical specialty training, 256–57; core medical colleges, 245–46, 252; evaluation training centers, 255; management training, 256; medical colleges research support, 246; schistosomiasis, 139; specialized training, 256
Ministry of Public Health, Division of Medical Education and Science: curriculum, 245; manpower plans, 245; teaching materials, 245
Monod, Jacques, genetics, 164
Morgan, T. H., Columbia University, 10, 41; genetics networks, 5; genetics, 8, 156, 158; taxonomy, 9
Mortality, causes, 239, *Tab.*, 241
Muller, H. J.: geneticist, 158; helping scientists, 161

N

Nakayama Shigeru: geology, 84; on Chidanken, 79
Nanking Higher Normal College, 44
Nanking, University of, 41; botanical research, 48; crop improvement research, 25; exchange program with Cornell University, 5, 23–24; genetics network with Cornell University, 5; herbarium, 44; plant genetics, 3, 9, 23–28; Szechwan Agriculture Improvement Institute, Chengtu, 26–28
National Academy of Peiping (later Chinese Academy of Sciences), 70; Geological Survey, 70
National Central University in Nanking, 44, 55; botanical research, 48; herbarium, 44; specimens sent to Arnold Arboretum, 60
National Coordinating Group on Male Antifertility Agents, 114; gossypol clinical trials, 115
National Economic Council, mineral resources, 87
National Institute for the Control of Pharmaceutical and Biological Products: JE vaccines, 148; vaccine efficiency, 153; viral vaccine program, 149

Nationalism, 92–95
Nationalist government, 94
National Peking University
  herbarium, 45
National Resources
  Commission, mineral
  resources, 87
National Southeastern
  University, 10–11; taxonomy,
  11
Naturalistic biology, 6–9
Naturalists, 92
Needham, J. G.: Cornell
  University naturalist, 7
Needham, Joseph: University of
  Cambridge historian of
  science, on geology, 87; on
  Ch'en Tze-ying, 17; on
  Szechwan Agriculture
  Improvement Institute, 28
New York Botanical Garden,
  43; exchange relations, 51
NORPLANT®, 113
Nung-pao, 26
Nurses, 243
Nutrient availability, *Fig.*, 212;
  increase, 210, 213
Nutritional status:
  anthropometric data, 220;
  child health, 223; food
  availability, 222; health, 222;
  improvement, 218–23; land
  reform, 222–23; micronutrient
  deficiency diseases, 226–27;
  rural-urban differences,
  223–26; sources of
  differences, 224–26; trends,
  219, 222

O

Oksenberg, Michel, xix
Olmsted, Frederick Law,
  Arnold Arboretum, 51
Oral history, xviii–xix
Ore formation, 77
Orleans, Leo A., xix
Orthodoxy, 91

P

Paleobotany, 80
Paleontologia Sinica, 80
Paleontology, Choukoutien, 82
Parasites, beneficial,
  Trichogramma spp., 188
Parathion, 193
Peach, 173
Peanut, 180–81
Pediatrics, undergraduate
  training, 248
Pei-p'iao (mining company),
  Geological Survey funds, 69
Pei-ta. See Peking University
P'ei Wen-chung, Choukoutien,
  82
Peking Man, 81–82
Peking Union Medical College,
  5, 242; Choukoutien, 82;
  Yenching University, 14–15.
  See also Capital Medical
  College of China
Peking University, 67; geology
  department, 68

People's Republic of China: Li Ssu-kuang, 86; population policy, 100; State Family Planning Commission local manufacture of Copper T devices, 109
Perkins, Dwight, agriculture statistics, 174
Persimmon, Japanese, 173
"Pesticide treadmill," 193
Pest management, migratory locusts, *Fig.*, 189
Pharmacists' assistants, 243
Pharmacy, Western, undergraduate programs, 249
Philippines Bureau of Science, 43
Physiologists, Rockefeller Foundation support, 41
Pi-Chao Chen, xix
Plant genetics, 23–28
Plant protection, 188; service, 195
Planthopper, 193–94
Polio, 142–44; Chinese oral vaccine, 143
Population Council of New York, 113; IUDs, 107
Population: growth, 174; policy, 99–100
Potato, 180
Praziquantel (pyquiton), 139
Predators, 194; *Coccinella septempunctata*, 188
Pregnancy, outcomes of, *Fig.*, 106
Public health, 243; changing practice, 248; faculties, 248

PUMC. *See* Peking Union Medical College
Pyquiton. *See* Praziquantel

Q

Qian Xinzhong, on contraception, 103

R

Race, in genetics, 21
Radiographers, 243
Reardon-Anderson, James, historian, xvii, xix; Kuomintang and science, 94; science in Republican China as international, 93
Rehder, Alfred, Arnold Arboretum herbarium curator, 53
Reisner, John H., University of Nanking, competition for collecting funds, 61
Republican China: biological sciences, 9; economic geology, 87; foreign support of science, 93–94; geology, 65, 67, 68; influences on science, 92
Republican period, naturalists in biology, 6–9
Research Institute of Geology, Academia Sinica, 65, 70; geophysics, 86; research, 68; structural geology, 85

Rhoades, Marcus, geneticist, 159
Rice (*Oryza sativa*), 171, *Tab.*, 173; Champa strains, 180; double cropping, 180; early-maturing varieties, 180, 185; hybrids, 185; indica varieties, 181; stem borer (*Chilo suppressalis*), 192, 194; virus diseases (yellow dwarf, yellow stunt, common dwarf), 193
Robbins, F. C., polio tissue cultures, 142
Rock, Joseph: competitive Western institutions, 61; plant collector, 52
Rockefeller Foundation, 5, 40; botanical institutions, 46; Cenozoic Research Laboratory, 82; China Program, 25; Cornell-Nanking program, support for, 24; education, 237; genetics, 4; graduate students, support for, 18; not research funds, 46–47; premedical education, 46; scholarships' development role, 17; T'an Chia-chen, support for, 20, 21, 22–23; taxonomy, 7; Tsinghua University, 11–12; vertebrate paleontology, 81
Roller, rice leaf (*Cnaphlocrocis medinalis*), 194
Royal Botanic Gardens, Kew, exchange relations, 51; systematic photography, 36
Royal Botanic Gardens, Edinburgh, exchange relations, 51

S

Sabin, Albert B., polio vaccine, 143
Sargent, Charles Sprague, 32, 51; Arnold Arboretum specimens policy, 60; Ch'en Huan-yung, plant collector, 54; Hu Hsien-su, specimen exchange, 55
Schistosoma japonicum, 125
Schistosomiasis: acute, 138; animal reservoirs, 131; articles, *Fig.*, 129; cattle treatment, 131; chronology and control, *Tab.*, 126; clinical study, 132, 133; control, 139; control assessment, 131–32; distribution, 127, 130, *Fig.*, 134–35; drug praziquantel (pyquiton), 137, 139; eggs, 132; eradication, 124–25, 138; eradication assessment, 131; history, 125–28; importance, 137; infections, *Fig.*, 136; intense study, 128; localized areas, 138; mass chemotherapy, 130; molluscicides, 137; prevention, 128; snail control, 131; treatment, 133; troops never exposed, 138

Schubert, Bernice, Harvard University, systematic photography, 36
Science: Chinese and Western integration, 93; missionary movement, 94; nationalistic politics, 94; role in China, 95
"Science and Life-view," xvii
Science Society, Biological Research Laboratory in Nanking, 11; China Foundation funding, 51; emphasis on Chinese conditions, 93; taxonomy, 7
Science Society of China, 11
Scientific and Technological Commission, 196, 198
SEU. *See* National Southeastern University
Shang-chih Hsüeh-hui. *See* "Aspiration Society"
Shanghai, 138
Shanghai Academy of Agricultural Sciences, 197
Shanghai First Medical College: educational development, 252; evaluation training, 255
Shao Baoruo, Institute of Parasitic Diseases, schistosomiasis control, 130
Shen Guangjing, locust control, 191
Shou-tu i-k'o ta-hsueh. *See* Capital Medical College of China
Sichuan. *See also* Szechwan
Sichuan Medical College, evaluation training, 255.

Sino-Japanese War: and research, 62–63; disrupts botanical research, 32
Smith, W. W., identification of botanical materials in Edinburgh, 62
Snail vector, 128
Sodium antimony tartrate, 130
Society for Corporate Research in Earth Science. *See* Chidanken
Soochow University: genetics networks, 5–6; taxonomy, 17–18; Yenching University connection, 18
Soybean, 173
Spiders, 192, 194
Spiess, Eliot, on Li Zhongjun, 162
Ssu (Sze) Hsing-chien, paleobotanist, 80
Stadler, L. A., University of Missouri, maize genetics, 159
State Family Planning Commission, 113
STC. *See* Scientific and Technological Commission
Steroidal contraceptives: injectable, 112; impact, 111; dosage, 111; research, 109–13; subdermal implant, 113; "vacation pill," 112
Stevens, Nettie, chromosome theory, 10
Steward, Albert Newton: Arnold Arboretum, 58; botanist, 42; competition for collecting funds, 61; need for

support, 42; University of
Nanking herbarium, 44–45
Stink bug, 192
Stomatology: assistants, 243;
undergraduate program, 249
Structural geology, 85–87, 89
Sturtevant, A. H.: California
Institute of Technology
geneticist, 157, 158; and T'an
Chia-chen, 21
Su Delong, Shanghai First
Medical College,
schistosomiasis, 124
Sun Yün-chu, invertebrate
paleontologist, 81
Sweet potato, 180
Symposium on Chinese Culture,
73n. 13, 92
Szechwan. *See also* Sichuan
Szechwan Agriculture
Improvement Institute, 26

T

T'an Chia-chen, 156–58;
California Institute of
Technology, 156–57; Chekiang
University science faculty, 22;
evolution, 161; fruit flies, 21,
157; Fudan University, 158;
genetic character of race, 21;
genetics, 20–23; ladybird
beetles, 9, 20–21, 22–23,
156–58; on Lysenkoism, 29;
political skill, 163; population
genetics, 157; radiation
effects, 157, 162

Tan Jiazhen. *See* T'an
Chia-chen
Tanner, J. M., growth rate
data, 220–22
Taro, 173
Taxonomy, 91; centrality in
Chinese biology, 6, 91–92
Tea, 173
Technicians (medical), 243
Tegengren, F. R.: iron
resources, 75; mining advisor
to Geological Survey, 69
Temin, H. M.: on basic and
applied science, 169; on
genetics in China, 163;
reverse transcriptase, 164
TFR. *See* Total fertility rate
Tianjin, IUD study, 107, 109
Ti-chih Yen-chiu-so. *See*
Geological Institute
Tien Ch'i-chün, invertebrate
paleontologist, 80
Ting, V. K. *See* Ting
Wen-chiang
Ting Wen-chiang: China
Foundation Board, 50;
Geological Institute, 67;
Geological Survey of China
head, 68; geologist, 65;
mining industry, 73–74;
naturalist, 7; on funding, 69;
Peking University geology
department, 68
Tisdale, W. E.: evaluation of
Ch'en Chen's biology
department, 12; on T'an
Chia-chen, 20–21; Rockefeller

Foundation, 47–48, 49;
taxonomy, 7
Torrey, John, Columbia University, 36
Total fertility rate, 100
Traditional medicine, 242, 243; drugs, 254; evaluation, 255; expansion, 249; faculties, 248–49
Translation, geological books, 66–67
Treaty of Nanking (1842), 37
Treaty of Tientsin (1858), 37
Trichogramma, 192
Ts'ai Yuan-p'ei: Peking University geology department, 67; scientific research, 70
Tsinghua scholarships, 39
Tsinghua University, 11, 12, 39, geology department, 68
Tung Chang, Geological Survey of China research fellow, 69
Turnip, 173

U

UNFPA. *See* United Nations Fund for Population Activities
United Nations Fund for Population Activities, Country program, 109
United States, arable land, 174
Universities, role in agricultural research, 200. *See also* Higher education

U.S. Department of Agriculture, Bureau of Plant Industry, plant collecting in China, 38
U.S. National Herbarium, The Smithsonian Institution, 51
U.S. National Institutes of Health, on gossypol, 114

V

Vaccination, neutralizing antibody response, *Tab.*, 151
Vaccines: efficacy, *Tab.*, 152; local production, 142. *See also* Japanese encephalitis
Vavilov, Nicolai, plant geneticist, 160
Vegetarian nutrition, 184
Veitch & Sons, collection of Chinese flora, 37
Vertebrate paleontology, 81–82
Virology, Chinese technical competence, 143

W

Walker, Egbert H., botanist, 42
Walnut, 173
Wang Chu-chuan, Geological Survey of China, 75
Wang Ch'ung-yu, mineral deposits, 77
Wang Shou, University of Nanking classical

geneticist, 25
Water chestnut, 173
Weaver, Warren: reductionist philosophy, 48; Rockefeller Foundation, 48, 49
Weller, T. H., polio tissue cultures, 142
Weng Wen-hao: Geological Institute, 67; Geological Survey of China, 71; on theoretical and applied geology, 73; ore formation, 77
Western Academy of Sciences, geological study, 71
Wheeler, William Morton, Bussey Institution for Research on Applied Biology, intercession for Ch'en Huan-yung, 54
Wilson, E. B.: classical genetics, 10; Columbia University cytologist, 158
Wilson, E. H., Veitch & Sons plant collector, 52
Winter melon, 173
Wong, W. H. *See* Weng Wen-hao
Woon-young Chun. *See* Ch'en Huan-yung
World Bank: agricultural education and research, 235; Chinese higher education, 233; education funds, 236, 237; instructional materials, 236; laboratory equipment, 234; medical education, 258; secondary vocational schools, 236; staff development, 234; teacher education, 236
World Health Assembly global attack on smallpox, 141
World Health Organization, 112
Wright, Sewall, population genetics, 159
Wu Ch'i-chün, botanical philologist, 57
Wu Chin-fu: published with Boring, 18; Soochow University biology department head, 17; Yenching University biology department, 17
Wu Mi, cultural conservative, 56

Y

Yam, 173
Yang Chung-chien, paleontologist, 79; Choukoutien, 82; geology, 88, 89
Yenching University, 41; biology department, 15–20; genetics, 3, 5–6, 14–20; Soochow University, link with, 15
Yin Tsan-hsün, invertebrate paleontologist, 81
Yuan Fu-li, Geological Survey of China research fellow, 69
Yu Chien-chang, invertebrate paleontologist, 81

Yüeh Sen-hsün

Yüeh Sen-hsün, invertebrate paleontologist, 81

Z

Zdansky, Otto, vertebrate paleontology, 81
Zhongshan Medical College, educational development, 252
Zhongshan University Institute of Ecological Entomology, insect pest applied research, 200
Zhong Shaohua, xviii
Zhou Chengyao, geneticist, Zhejiang Agricultural University in Hangzhou, 159
Zhou, C. Y. *See* Zhou Chengyao
Zhou Enlai: family planning, 99; Ota ring, 107

Printed and bound by CPI Group (UK) Ltd, Croydon, CR0 4YY
09/06/2025

14685677-0002